Textbooks in Telecommunication Engineering

Series Editor

Tarek S. El-Bawab, Professor and Dean of Engineering,
American University of Nigeria, Yola, Nigeria

Dr. Tarek S. El-Bawab, who spearheaded the movement to gain accreditation for the telecommunications major is the series editor for Textbooks in Telecommunications. Please contact him at telbawab@ieee.org if you have interest in contributing to this series.

The Textbooks in Telecommunications Series:

Telecommunications have evolved to embrace almost all aspects of our everyday life, including education, research, health care, business, banking, entertainment, space, remote sensing, meteorology, defense, homeland security, and social media, among others. With such progress in Telecom, it became evident that specialized telecommunication engineering education programs are necessary to accelerate the pace of advancement in this field. These programs will focus on network science and engineering; have curricula, labs, and textbooks of their own; and should prepare future engineers and researchers for several emerging challenges.

The IEEE Communications Society's Telecommunication Engineering Education (TEE) movement, led by Tarek S. El-Bawab, resulted in recognition of this field by the Accreditation Board for Engineering and Technology (ABET), November 1, 2014. The Springer's Series Textbooks in Telecommunication Engineering capitalizes on this milestone, and aims at designing, developing, and promoting high-quality textbooks to fulfill the teaching and research needs of this discipline, and those of related university curricula. The goal is to do so at both the undergraduate and graduate levels, and globally. The new series will supplement today's literature with modern and innovative telecommunication engineering textbooks and will make inroads in areas of network science and engineering where textbooks have been largely missing. The series aims at producing high-quality volumes featuring interactive content; innovative presentation media; classroom materials for students and professors; and dedicated websites.

Book proposals are solicited in all topics of telecommunication engineering including, but not limited to: network architecture and protocols; traffic engineering; telecommunication signaling and control; network availability, reliability, protection, and restoration; network management; network security; network design, measurements, and modeling; broadband access; MSO/cable networks; VoIP and IPTV; transmission media and systems; switching and routing (from legacy to next-generation paradigms); telecommunication software; wireless communication systems; wireless, cellular and personal networks; satellite and space communications and networks; optical communications and networks; free-space optical communications; cognitive communications and networks; green communications and networks; heterogeneous networks; dynamic networks; storage networks; ad hoc and sensor networks; social networks; software defined networks; interactive and multimedia communications and networks; network applications and services; e-health; e-business; big data; Internet of things; telecom economics and business; telecom regulation and standardization; and telecommunication labs of all kinds. Proposals of interest should suggest textbooks that can be used to design university courses, either in full or in part. They should focus on recent advances in the field while capturing legacy principles that are necessary for students to understand the bases of the discipline and appreciate its evolution trends. Books in this series will provide high-quality illustrations, examples, problems and case studies.

For further information, please contact: Dr. Tarek S. El-Bawab, Series Editor, Professor and Dean of Engineering, American University of Nigeria, telbawab@ieee.org; or Mary James, Senior Editor, Springer, mary.james@springer.com

More information about this series at http://www.springer.com/series/13835

Mubashir Husain Rehmani

Blockchain Systems and Communication Networks: From Concepts to Implementation

 Springer

Mubashir Husain Rehmani
Department of Computer Science
Munster Technological University (MTU)
Cork, Ireland

Additional material to this book can be downloaded form https://www.springer.com/book/9783030717872

ISSN 2524-4345   ISSN 2524-4353   (electronic)
Textbooks in Telecommunication Engineering
ISBN 978-3-030-71790-2   ISBN 978-3-030-71788-9   (eBook)
https://doi.org/10.1007/978-3-030-71788-9

This Springer imprint is published by the registered company Springer Nature Switzerland AG
The registered company address is: Gewerbestrasse 11, 6330 Cham, Switzerland

This book is devoted to my dearest Sheikh, Grandmother, Father, Mother, and Brother!

Preface

Internet has been used to share information among different parties. For instance, customers make online transactions in banks, online buying and selling, management of digital currencies, and financial transactions are few examples where information is shared among different parties. Traditional way of doing these transactions requires the presence of the trusted third party. Blockchain, in the absence of this trusted third party, permits communicating parties to interact with each other. Blockchain is basically a distributed and decentralized public ledger system used for maintaining the transactions record over several computers (blockchain nodes). In fact, Distributed Ledger Technology (DLT) ensures the availability of multiple copies of the identical ledger distributed across various places. If any change happens in any place in the ledger, it will be reflected in all the places.

Blockchain has been applied to numerous applications areas ranging from health sector to transportation and from financial sector to energy management systems. This wide applicability of blockchain technology is due to its inherent features like decentralization, auditability, and fault tolerance. Blockchain can play a vital role in communication networks as well. Let's take an example of Internet of Things (IoT). In IoT, blockchain can be used for a decentralized fabric for the IoT, with no managing or authorizing intermediaries. Similarly, blockchain can also provide IoT identity and data management, privacy, trustless architectures and secured communications, and monetization of IoT data and resources.

Considering the aforementioned applications and the importance of this topic, I have been working on this topic with my research collaborators and Ph.D. students since January 2018. In order to equip myself fully with the advent of this technology, I tried to take different online courses, attended several webinars, and read several books on this topic. In addition to this, without exaggeration, I read hundreds of research papers on this so-called disruptive technology blockchain. Fortunately, I had been given a chance to design two modules on distributed ledger technology. The first one is for the undergraduate programs on blockchain and the second one is on distributed ledger technology for graduate programs, both at the Department of Computer Science, Munster Technological University (MTU), Ireland. The module distributed ledger technology had to be delivered to programs such as Masters in Artificial Intelligence, Masters in Cloud Computing, and Masters in Cybersecurity.

Both proposed modules were accepted and became the part of the curriculum at MTU.

I wanted to design a module which not only provides solid theoretical background to students but also enables them to easily think about blockchain realization and its applications. More precisely, questions like: how we can adopt any specific blockchain architecture to a specified application and which consensus algorithm can be used in each application scenario? What are the security and privacy concerns associated with each type of blockchain design? All these questions were dispersed and can be found in different resources such as books and research papers. However, there was not a single source which completely answers all these questions to the extent in which I was searching. On top of it, the main aspect which I was looking for was the coding aspect–in order to give real sense of using blockchain to students.

Since the module that I designed has five credit hours, i.e., it has 2 hours of lab in each week (integral part of this module), therefore, I realized that not a single easy-to-use resource or book is available that helps students to understand the working of blockchain. For instance, how hashing can be implemented? What will be the impact if blocks get tampered by anyone? How we can implement different consensus algorithms? How blocks are validated and broadcast? All such questions spanning from theoretical concepts to their implementation were not available in a single source so that one can easily understand this blockchain technology and easily implement the ideas presented therein by using an open source programming language. Moreover, a textbook on applying blockchain technology for communication systems is missing. Therefore, considering this gap, I was motivated enough to think about writing a textbook on blockchain technology which not only provides theoretical knowledge to the students but also helps them understand basic ideas by implementing them.

I would like to thank *Muneeb Ul Hassan* who helped me in the preparation of lab material for the above modules, which I then used as a basis to explain blockchain concepts from the implementation perspective in this book. Without the help of *Muneeb Ul Hassan*, I may not be able to produce such an easy and understandable source code. Finally, I would like to thank *Prof. Tarek El-Bawab*, who invited me and gave me the opportunity to publish this book under *Textbooks in Telecommunication Engineering* by Springer.

I would like to say my special thanks to *Tim Horgan*—Head of Faculty of Engineering and Science at Munster Technological University (MTU) and *Donna O'Shea*—Chair Cybersecurity and the former Head of Department of Computer Science at MTU. I remember, we all were taking tea together after a meeting and there Tim and Donna suggested me to prepare a module on blockchain technology. This was the time when I seriously started thinking about writing a textbook on blockchain technology.

This book is particularly written for the Computer Science and Telecom students. This book in fact can serve as a step-by-step hands-on tutorial for designing and implementing blockchain systems besides building concrete blockchain theoretical knowledge. To support further reading, few interesting things have been included in each chapter: further reading section (what to do next?), research directions, basic definitions, programming tips, labs, and self-assessment exercises.

The objective of this book is to provide detailed insights on blockchain systems, starting from its historical perspective and moving toward building foundational knowledge about blockchain systems. This book also covers blockchain systems with emphasis on applications to implementation considering Communication Networks and Services, rather than books which only covers either blockchain architectures, cryptocurrencies, or about building blockchain projects. This book also discusses the technologies related to the integration of telecommunication systems and distributed ledger technology (blockchain). This book bridges the divide between the fields of telecommunication networks (including computer and mobile networks) and blockchain systems, while focusing on the applicability of blockchain in different applications domains and its implementation.

This book is organized into three parts:

- Part I: "Blockchain Systems: Background, Fundamentals, and Applications"
- Part II: "Hands-on Exercises and Blockchain Implementation"
- Part III: "Blockchain Systems and Communication Networks"

Part I: "Blockchain Systems: Background, Fundamentals, and Applications" consists of four chapters. In Chap. 1, blockchain introduction is provided. Chapter 2 discusses the differences between database management system and blockchain. Blockchain fundamentals and working principles are discussed in Chap. 3 and finally, Chap. 4 is dedicated to consensus algorithms in blockchain systems. Part II: "Hands-on Exercises and Blockchain Implementation" consists of one chapter (Chap. 5) in which two mini projects are presented. Moreover, this chapter also contains five lab implementations along with desired program output and sample code. Finally, in Part III: "Blockchain Systems and Communication Networks", two chapters are included. The first chapter (Chap. 6) discusses cognitive radio networks and blockchain. The second chapter (Chap. 7) talks about communication networks and blockchain in general covering various communication networks such as Wi-Fi, cellular networks, cloud computing, Internet of Things, software defined network, and smart energy networks.

I hope you will enjoy reading this book and find it beneficial, particularly from hands-on exercises and the implementation point of view.

Cork, Ireland   Mubashir Husain Rehmani
February 2021

Acknowledgements

I would like to express sincere thanks to Allah Subhanahu Wa-ta'ala that by his grace and bounty I was able to write this book.

I wish to express gratitude to Sheikh Hazrat Mufti Mohammad Naeem Memon Sahib Damat Barakatuhum of Hyderabad, Pakistan. I could have not written this book without his prayers, spiritual guidance, and moral support.

I also want to acknowledge my family, especially my wife, for her continued support and encouraging words that helped me to complete this book.

Last but not least, I also want to thanks Saad, Maria, and Aamir for their patience and support during the write-up of this book.

Contents

Part I   Blockchain Systems: Background, Fundamentals, and Applications

1   Introduction to Blockchain Systems 3
   1.1   From Ledger to Distributed Ledger Technologies 3
       1.1.1   Classification of Distributed Ledger Technology 4
       1.1.2   Blockchain 5
       1.1.3   Directed Acyclic Graph (DAG) 7
   1.2   Features of Blockchain Systems 8
       1.2.1   Decentralization 8
       1.2.2   Transparency 9
       1.2.3   Immutability 9
       1.2.4   Availability 9
       1.2.5   Pseudonymity 9
       1.2.6   Security 9
       1.2.7   Non-Repudiation 9
       1.2.8   Auditability 10
       1.2.9   Data Tampering 10
   1.3   A Great Example for the Use of Blockchain Technology:
       Food Supply Chain 11
       1.3.1   Traceability and Provenance Within Food Supply
               Chain 11
       1.3.2   Identification and Removal of Contaminated Food 12
       1.3.3   Blockchain for Food Supply Chain 12
   1.4   Summary 13
   1.5   Further Reading 13
       1.5.1   General Blockchain History and Background 13
       1.5.2   Food Supply Chain and Blockchain 13
   Problems 14

2 Blockchain Technology and Database Management System 15
  2.1   Distributed Ledger Technology and Database Management
       System 15
  2.2   When to Select Blockchain Over DBMS? 17
  2.3   Blockchain and Database Maintenance 18
      2.3.1   Ledger Maintenance in Public Blockchain 18
      2.3.2   Ledger Maintenance in Consortium Blockchain 18
      2.3.3   Ledger Maintenance in Private Blockchain 18
  2.4   Database System, DLT, and Public Verifiability 19
  2.5   Comparison of Blockchain Systems and Traditional DBMS 19
  2.6   Large-Scale Distributed Database Systems and Blockchain 20
  2.7   Trust and Public Availability of Blockchain 21
  2.8   How Blockchain Is Different from Distributed Data Storage? 21
  2.9   Summary 21
  2.10 Further Reading 21
  Problems 22

3 Blockchain Fundamentals and Working Principles 23
  3.1   Blockchain Network 23
      3.1.1   Public Blockchain Network—Permissionless 24
      3.1.2   Private Blockchain Network—Permissioned 24
      3.1.3   Consortium Blockchain Network—Permissioned 25
  3.2   General Issues with Public Blockchain 25
      3.2.1   Limited Transactions 26
      3.2.2   Scalability 26
      3.2.3   Pseudonymity 26
      3.2.4   Block Size 26
      3.2.5   Energy Consumption 26
  3.3   Underlying Network for Peer Discovery and Topology
       Maintenance in Blockchain 27
  3.4   Broadcasting in Blockchain Network 27
  3.5   Users/Nodes in a Blockchain Network 27
      3.5.1   Full Blockchain Nodes 28
      3.5.2   Lightweight Blockchain Nodes 28
      3.5.3   Miner Nodes 28
  3.6   Blockchain Nodes as Leaders and Validators 29
  3.7   Blockchain Nodes as Sender and Receiver 29
  3.8   Layers in Blockchain 30
      3.8.1   Application Layer 30
      3.8.2   Virtualization and Smart Contract Layer 31
      3.8.3   Consensus Layer 31
      3.8.4   Network and OS Layer 31
      3.8.5   Data Organization and Topology Layer 32
      3.8.6   Hardware Layer 32
  3.9   General Working Sequence of Blockchain 32

	3.9.1	Transaction	33
	3.9.2	Transaction Signing	34
	3.9.3	Transaction Verification	34
	3.9.4	Transaction Broadcast	34
	3.9.5	Transaction/Block Validation	34
	3.9.6	Block Confirmation	34
3.10	Composition of a Block		35
	3.10.1	Hash Pointer	35
	3.10.2	Merkle Tree	35
3.11	Blockchain Governance System: Who Owns Blockchain?		36
3.12	Who Make Modifications in Blockchain?		36
3.13	Confidentiality in Blockchain		36
3.14	Blockchain Platforms		37
	3.14.1	Availability of Blockchain Platforms	37
	3.14.2	Blockchain Platform Suitable only for Cryptocurrency	38
	3.14.3	Blockchain Platform that Supports Smart Contracts (Business Logic)	38
	3.14.4	Blockchain Platform Available over the Cloud	38
3.15	Blockchain as a Service (BaaS)		38
3.16	BitCoin Blockchain		39
	3.16.1	Creating Trust in Bitcoin Blockchain	39
	3.16.2	Working of Bitcoin	40
3.17	Ethereum Blockchain		41
3.18	Hyperledger		42
3.19	Corda		42
3.20	Tendermint		43
3.21	Chain Core		44
3.22	Quorum		44
3.23	Key Generation and Blockchain Digital Signature Procedure		45
3.24	Data Models in Blockchain		45
3.25	Implementation and Performance Evaluation Tools for DLTs		45
	3.25.1	Hyperledger Caliper	46
	3.25.2	BlockBench	46
	3.25.3	DAGBench	47
	3.25.4	How Consensus Algorithm Can Impact on the Performance of Blockchain?	47
3.26	Hashing in Blockchain		47
	3.26.1	Hashing Applied to Ethereum Blockchain	48
3.27	Data Storage in Blockchain		48
3.28	Data Structure in Blockchain		49
3.29	Privacy of Nodes in Blockchain		49
3.30	Smart Contracts		49
	3.30.1	Ethereum	49
	3.30.2	Hyperledger	50

3.30.3   Tendermint 50
3.30.4   Energy Web Chain (EW Chain) 50
3.31   Scalability Issues in Blockchain Systems 50
3.31.1   Blockchain Scalability Issues and Communication
Networks 51
3.32   How to Increase the Transaction Capacity of Blockchain
Systems? 51
3.32.1   Off-Chain Transactions 52
3.32.2   Sharding 52
3.33   Interoperability in Blockchain Systems 52
3.33.1   Example to Understand Interoperability Issue 53
3.33.2   Using Smart Contract for Interoperability 53
3.33.3   Using Exchange for Interoperability 54
3.33.4   Consensus Protocols and Interoperability Issue 54
3.33.5   Interoperability Between Old and New Blockchain
Systems 54
3.33.6   Transaction Speed and Interoperability 55
3.33.7   Semantic and Syntatic Interoperability 55
3.33.8   Transaction Fees and Interoperability 55
3.33.9   Tokens and Interoperability 55
3.34   Summary 56
3.35   Future Research Direction 56
3.36   Further Reading 57
Problems 57

4   Blockchain Consensus Algorithms 61
4.1   Consensus Algorithms 61
4.2   Functionality of Consensus Algorithm 62
4.3   Proof-of-Work (PoW) Consensus Algorithm 63
4.3.1   Leader Node 65
4.3.2   Issues in PoW 66
4.3.3   How PoW Deals with Attacks? 66
4.3.4   Example of PoW Consensus Algorithm 66
4.4   Proof of Stake (PoS) Consensus Algorithm 67
4.4.1   Issues in PoS 68
4.5   Mining Pools 68
4.6   Issues Related with Mining Pools 70
4.7   Transaction (Tx) Throughput 70
4.8   Block Confirmation Time 71
4.9   Impact of Tx Throughput and Block Size 71
4.10   Impact of Block Confirmation Time and Throughput 71
4.11   Impact of Transaction Size and Throughput 72
4.12   Example of Tx Throughput and Block Confirmation Time 73
4.13   Different Consensus Algorithms 73
4.13.1   Proof-of-X 73

4.13.2 Hyrid Consensus Protocol 73
4.13.3 PoW-PoS Protocols 74
4.13.4 Committee-Based Consensus Algorithms 74
4.13.5 Consensus Protocols for Distributed Data Storage 74
4.13.6 Proof-of-Human-Work 74
4.13.7 Primecoin 75
4.13.8 Proof-of-Exercise 75
4.13.9 Proof-of-Useful-Work 75
4.13.10 Ouroboros Conesus Protocol 75
4.13.11 Chain of Activity 76
4.13.12 Casper 76
4.13.13 Algorand 76
4.13.14 Tendermint 77
4.14 Consensus Protocol for Permissioned Blockchain 77
4.15 Consensus Protocol for Permissionless Blockchain 77
4.16 Why BFT Protocols Cannot Be Used in Public Blockchain? 77
4.17 Summary 78
4.18 Further Reading 78
Problems 78

Part II   Hands-on Exercises and Blockchain Implementation

5   Hands-On Exercise and Implementation 81
5.1 Mini Project 1: Critical Analysis of Distributed Ledger
      Technology 81
      5.1.1 Questions 82
5.2 Mini Project 2: Implementation of Distributed Ledger
      Technology and It's Security Analysis 83
      5.2.1 Questions 84
5.3 Lab Implementation 1 85
      5.3.1 Aim 85
      5.3.2 Steps to Follow 86
      5.3.3 Desired Program Output 86
      5.3.4 Sample Code 86
5.4 Lab Implementation 2 88
      5.4.1 Aim 88
      5.4.2 Steps to Follow 89
      5.4.3 Desired Program Output 89
      5.4.4 Sample Code 89
5.5 Lab Implementation 3 91
      5.5.1 Aim 91
      5.5.2 Steps to Follow 91
      5.5.3 Desired Program Output 93
      5.5.4 Sample Code 93
5.6 Lab Implementation 4 96

5.6.1   Aim 96
5.6.2   Steps to Follow 96
5.7   Lab Implementation 5 96
5.7.1   Aim 96
5.7.2   Steps to Follow 96
5.8   Hands-On Exercise 97
5.8.1   Exploring Real Blockchain: Bitcoin 97
5.8.2   Exploring Real Blockchain: Ethereum 98
5.8.3   Exploring Real Blockchain: Bitcoin Cash: Fork
          of Bitcoin 99
5.8.4   Exploring Real Blockchain: Bitcoin Blocks
          Linkage 100
5.8.5   Exploring Real Blockchain: Bitcoin's UTXO
          Concept 100
5.8.6   Exploring Real Blockchain: Ethereum's Block
          Contents 100
5.8.7   How Many Byzantine Nodes (Faulty Nodes)
          a Blockchain Network Can Tolerate? 100
5.8.8   How to Find the Size of Ethereum Blockchain? 102
5.8.9   How to Find the Transaction Handling Capacity
          of Blockchain? 102
5.8.10   How to Find Tx Throughput and Block
          Confirmation Time 102
5.8.11   How to Find Wining Probability in PoW Consensus ... 102
5.9   Summary 103

Part III   Blockchain Systems and Communication Networks

6   Cognitive Radio Networks and Blockchain 107
6.1   Wired and Wireless Communication Systems 108
6.2   Dynamic Spectrum Access (DSA) 109
6.3   Blockchain and Spectrum Management 110
6.3.1   Time Granularity and its Exploitation for Spectrum
          Trading Through Blockchain 112
6.3.2   Use of Tokens in Dynamic Spectrum Management
          (DSM) 112
6.4   Usage of Blockchain Technology from the Spectrum
       Licensing Perspective 113
6.4.1   Licensed Spectrum Band 113
6.4.2   Shared Licensed Spectrum Band 113
6.4.3   Unlicensed Spectrum Band 114
6.5   Blockchain Enabled Cognitive Radio Network
       and Collision-Free Communication 114
6.5.1   Collision-Free Communication 114

	6.5.2	Blockchain-Enabled Cognitive Radio Network and CFC	115
6.6	Medium Access by CR Nodes as an Auction	116
6.7	Advantages of Using Blockchain Technology in Dynamic Spectrum Management (DSM)	117
	6.7.1	Lack of Central Entity	117
	6.7.2	Immutability	117
	6.7.3	Availability	117
	6.7.4	DoS Resilient	118
	6.7.5	Non-repudiation	118
	6.7.6	Smart Contract Integration	118
6.8	Spectrum Patrolling Through Blockchain	118
6.9	Issues and Challenges When Deploying Blockchain to Dynamic Spectrum Management	119
6.10	Summary	120
6.11	Future Research Directions	120
6.12	Further Reading	120
	6.12.1	Blockchain and Spectrum Management	121
Problems		121

7 Communication Networks and Blockchain 123
7.1	Blockchain and Internet of Things (IoT)	125
7.2	Blockchain for Fog-RAN	126
7.3	Blockchain and IoT Edge	126
	7.3.1	Challenges in Blockchain-Based IoT Edge	128
7.4	Blockchain, IoT, and Consumer Electronics	128
	7.4.1	How to Manage IoT and CE Massive Data?	129
	7.4.2	Which Blockchain to Use for CE and IoT Devices?	129
7.5	Blockchain and Wireless Power Transfer—Green IoT	130
7.6	Blockchain and Internet of Vehicles (IoV)	130
7.7	Blockchain, Software Defined Networks (SDN), and Virtualization	131
	7.7.1	Blockchain-Based SDN: Advantages	131
	7.7.2	Virtualization, Cloud Computing, Edge, and Fog Computing	131
7.8	Blockchain and Cloud of Things	132
7.9	Blockchain in Cellular Networks	133
	7.9.1	Blockchain and Mobile Devices	133
	7.9.2	Blockchain and Roaming in Cellular Networks	134
7.10	Blockchain and Wi-Fi Networks	134
7.11	Multimedia Communication Networks and Blockchain	134
	7.11.1	Video Streaming Communication Networks and Blockchain	135
	7.11.2	New Methods of Revenue Generation and Business Models	136

7.11.3 Auditing for Video Content Generated Revenue 136
7.11.4 Smart Contracts for Video Content 136
7.11.5 Peer-to-Peer Video Content Sharing 137
7.11.6 Resolving of Privacy Issues Through Blockchain 137
7.11.7 Fake Video Generation and Tracking 137
7.11.8 Privacy of Video Content 137
7.12 Smart Grid Communication System and Blockchain 137
7.12.1 Prosumers 137
7.12.2 Energy Trading Benefits 138
7.12.3 Privacy Preservation in Blockchain-Enabled
Smart Grid 139
7.12.4 Vehicle to Grid (V2G) Energy Trading 141
7.12.5 Effect of DoS on Energy Trading Market 142
7.12.6 Cryptocurrency in Energy Trading Systems 142
7.12.7 Arbitrage in Energy Trading Systems/Markets
Through Blockchain Systems 142
7.12.8 Renewable Energy Resources and Negative Pricing 143
7.13 Communication Networks and the Use of Blockchain
with Machine Learning 144
7.13.1 Machine Learning and Communication Networks 144
7.13.2 Classification of Machine Learning Techniques
and Blockchain 144
7.13.3 Advantages of Using Machine Learning
in Blockchain-Enabled Communications
Networks 145
7.14 Summary 146
7.15 Future Research Directions 146
7.15.1 Blockchain, Smart Grid, and Peer-to-Peer Energy
Trading 147
7.16 Further Reading 147
7.16.1 Blockchain and IoT, Edge, Fog, and Cloud
Computing 147
7.16.2 Blockchain, Wi-Fi, and Mobile Communication 148
7.16.3 Smart Grid and Blockchain 148
7.16.4 Multimedia and Blockchain 148
7.16.5 Blockchain, Machine Learning,
and Communication Networks 149
Problems 149

Solutions 151

References 155

Index 161

Information About the Author

Mubashir Husain Rehmani (M'14—SM'15)
received the B.E. degree in computer systems engineering from Mehran University of Engineering and Technology, Pakistan, in 2004, the M.S. degree from the University of Paris XI, Paris, France, in 2008, and the Ph.D. degree from the University Pierre and Marie Curie, Paris, in 2011. He is currently working as an Assistant Lecturer in Department of Computer Science at Munster Technological University (MTU), formerly known as Cork Institute of Technology (CIT), Ireland. He worked at Telecommunications Software and Systems Group (TSSG), Waterford Institute of Technology (WIT), Waterford, Ireland as Post-Doctoral researcher from September 2017 to October 2018. He served for 5 years as an Assistant Professor at COMSATS Institute of Information Technology, Wah Cantt., Pakistan. He is currently an Area Editor of Wireless Communications of the IEEE Communications Surveys and Tutorials. He served for 3 years (from 2015 to 2017) as an Associate Editor of the IEEE Communications Surveys and Tutorials. He is also serving as Column Editor for Book Reviews in IEEE Communications Magazine. He has been appointed as Editor in IEEE Transactions on Green Communications and Networking (TGCN). Currently, he serves as Associate Editor of *Journal of Network and Computer Applications* (Elsevier), and the Journal of Communications and Networks (JCN). He is also serving as a Guest Editor of *Ad Hoc Networks* (Elsevier), *Future Generation Computer Systems* (Elsevier), the IEEE Transactions on Industrial Informatics, and *Pervasive and Mobile Computing* (Elsevier).

He has authored/edited two books published by IGI Global, USA, two books published by CRC Press, USA, and one book with Wiley, U.K. He received "Best Researcher of the Year 2015 of COMSATS Wah" award in 2015. He received the certificate of appreciation, "Exemplary Editor of the IEEE Communications Surveys and Tutorials for the year 2015" from the IEEE Communications Society. He received Best Paper Award from IEEE ComSoc Technical Committee on Communications Systems Integration and Modeling (CSIM), in IEEE ICC 2017. He consecutively received research productivity award in 2016–17 and also ranked # 1 in all Engineering disciplines from Pakistan Council for Science and Technology (PCST), Government of Pakistan. He also received Best Paper Award in 2017 from Higher Education Commission (HEC), Government of Pakistan. He is the recipient of Best Paper Award in 2018 from Journal of Network and Computer Applications (Elsevier). Dr. Rehmani received $Highly\,Cited\,Researchers^{TM}$ awards 2020 by Clarivate Analytics ($Web\,of\,Science^{TM}$). His performance in this context features in the top 1% in the field of Computer Science. Dr. Rehmani is the only researcher from Ireland in the field of "Computer Science" who received this international prestigious award.

News Coverage in the Media

Dawn newspaper.
https://www.dawnnews.tv/news/1151426/
Jang newspaper—07th Dec 2020.
https://jang.com.pk/news/855049
The Cork, Ireland Newspaper—22nd Nov 2020.
https://www.thecork.ie/2020/11/22/cork-lecturer-ran
ked-in-top-1-of-most-influential-researchers-worldw
ide-in-the-field-of-computer-science/
The Express Tribune—05th Dec 2020.
https://tribune.com.pk/story/2274782/pakistani-lec
turer-ranked-among-worlds-top-1-computer-science-
researchers
SAMAA TV English News—01 Dec 2020.
https://www.samaa.tv/news/2020/12/pakistani-lec
turer-named-among-worlds-top-computer-science-res
earchers/

Silicon Republic Website, Ireland—18th Nov 2020.
https://www.siliconrepublic.com/innovation/ireland-
elite-scientists-2020
The Echo Newspaper, Ireland—11th Feb 2021.
https://www.echolive.ie/corknews/arid-40224959.html
SAMAA TV Naya Din Morning Show—1st Dec 2020.
https://www.youtube.com/watch?v=Xuq0daxtzko
Express News Expresso Morning Show—7th Dec 2020.
https://www.youtube.com/watch?v=Z3QTkSmZRyE
ARY News Channel—9 pm, 5th Dec 2020. Watch from
minute 18:30.
https://videos.arynews.tv/arynews-bulletin-9-pm-5-dec
ember-2020/

Acronyms

ADSL	Asymmetric Digital Subscriber Line
B5G	Beyond 5G
BaaS	Blockchain As A Service
BFT	Byzantine Fault Tolerant
BMaaS	Bitcoin Mining As A Service
BTS	Base Transceiver Station
CA	Contract Account
CAPTCHA	Computers and Humans Apart
CBRS	Citizen Band Radio Service
CDMA	Code Division Multiple Access
CE	Consumer Electronics
CFC	Collision Free Communication
CoA	Chain of Activity
CR	Cognitive Radio
CRN	Cognitive Radio Network
CRSN	Cognitive Radio Sensor Network
D2D	Device to Device Communication Network
DAG	Directed Acyclic Graph
DB	Database
DBMS	Data Base Management System
DHT	Distributed Hash Table
DL	Distributed Ledger
DLT	Distributed Ledger Technology
DNS	Domain Name Server
DSA	Dynamic Spectrum Access
DSM	Dynamic Spectrum Management
ELN	Energy Local Network
EOA	Externally Owned Account
EVM	Ethereum Virtual Machine
FCC	Federal Communication Commission
FDMA	Frequency Division Multiple Access
FTS	Follow the Satoshi

GHz	Giga Hertz
GSM	Global System for Mobile Communication
Hash-DAG	Hash-based Directed Acyclic Graph
HDFS	Hadoop Distributed File System
HDSL	High-bit-rate Digital Subscriber Line
HNMO	Home Network Mobile Operator
IIoT	Industrial Internet of Things
IoT	Internet of Things
IoV	Internet of Vehicles
ISM	Industrial, Scientific, and Medical
JVM	Java Virtual Machine
KHz	Kilo Hertz
kNN	k-Nearest Neighbor
MAB	Multi-Armed Bandit
MAC	Medium Access Control
MANET	Mobile Ad Hoc Network
MG	Micro Grid
MHz	Mega Hertz
ML	Machine Learning
MNO	Mobile Network Operator
MS	Mobile Station
MT	Machine Type
NMG	Networked Micro Grid
OS	Operating System
P2P	Peer-to-Peer Network
PBFT	Practical Byzantine Fault Tolerant
PCA	Principal Component Analysis
PLC	Power Line Communication
PoA	Proof of Activity
PoET	Proof of Elapsed Time
PoN	Proof of Networking
PoS	Proof of Stake
PoW	Proof of Work
PU	Primary User
QoS	Quality of Service
RADIUS	Remote Authentication Dial-In User Service
RAN	Radio Access Network
RER	Renewable Energy Resources
RL	Reinforcement Learning
RSU	Road Side Unit
SaaS	Software As A Service
SAN	Storage Area Networks
SDH	Synchronous Digital Hierarchy
SDN	Software Defined Network
SG	Smart Grid

SHA	Secure Hash Algorithm
SONET	Synchronous Optical Networking
SQL	Structured Query Language
SU	Secondary User
SVM	Support Vector Machine
TDMA	Time Division Multiple Access
TES	Transactive Energy System
THz	Tera Hertz
TVWS	TV White Space
UTXO	Unspent Transaction Output
UWSN	Under Water Sensor Network
V2G	Vehicle to Grid
VANET	Vehicular Ad Hoc Network
VDSL	Very high-speed Digital Subscriber Line
VNMO	Visiting Network Mobile Operator
VP2P	Virtual P2P
VPP	Virtual Power Plant
WDM	Wavelength-division multiplexing
WSAN	Wireless Sensor and Actor Network

Part I
Blockchain Systems: Background, Fundamentals, and Applications

This part of the book provides background and fundamental knowledge about blockchain systems and consists of four chapters. In chapter one, blockchain introduction is provided. Chapter two discusses the differences between database management system and blockchain. Blockchain fundamentals and working principles are discussed in chapter three and finally, chapter four is dedicated to consensus algorithms in blockchain systems.

Chapter 1
Introduction to Blockchain Systems

1.1 From Ledger to Distributed Ledger Technologies

People keep assets in the form of money, land area, shops, vehicles, and agricultural land. These assets need to belong to a person, group of persons, or an entity such as business or government. Moreover, these assets can be sold and buy from one party to another. In this context, first the ownership has to be proved. Secondly, when transferring this property or assets from one party to another, the ownership record has to be updated. Similarly, when a person dies, all his/her assets have to be moved to the heirs. This also requires record maintenance of the property or assets ownership. This record maintenance is known as "ledger". A ledger can record asset transfer within an organization. For instance, a ledger for payroll, a ledger for bills, a ledger for amount received as income, and a ledger for amount payable. All these ledgers can be linked together to form a bigger ledger. Depending upon the type of record, ledgers can be classified into different categories. For instance, land record can be maintained in a ledger maintained by the land management department. Business can track their buying and selling, and this can be recorded in general ledger or account ledger.

From centuries, this has been done manually and the record has been saved on printed registers, thus a whole department has been established and named as Registry. Patwari is another similar term used for record keeping (agriculture land, corps production, etc.) at village level, and still in place in Indian sub-continent. Though some efforts have been made to replace this old Patwari system in Punjab, Pakistan but still it is in place. Even in the developing countries, we can find that this record keeping for the management of assets is still happening on papers and registers.

From the last few decades, this practice has been changed and this record keeping in ledgers is done through computers, thus we had these digital ledgers. In the beginning, these digital ledgers were maintained using word processing software. As soon as more structured record maintenance was required, ledger management was moved from word processing software to spreadsheets. However, these spreadsheets still cannot handle million of records and querying them and extract useful informa-

© Springer Nature Switzerland AG 2021
M. H. Rehmani, *Blockchain Systems and Communication Networks: From Concepts to Implementation*, Textbooks in Telecommunication Engineering,
https://doi.org/10.1007/978-3-030-71788-9_1

Table 1.1 The evolution of ledger technology

Ledger technology	Description
Ledgers	Record kept on registers and printed books and managed manually
Digital Ledgers	Record kept on computer using software such as word processing or spreadsheets
Distributed Digital Ledgers	Record kept on several computers, but a central entity manages it
Decentralized Distributed Digital Ledgers	Record kept on several computers but managed in a decentralized manner. This is also known as blockchain technology

tion was quite complex. Additionally, these spreadsheets are not tampering resistant. Thus, database technology evolved which served the purpose of digital ledger very well.

> A "Distributed Ledger (DL)" is basically a ledger that keeps digital data, synchronized and shared over several machines (nodes) in a geographically distant locations, without administering these machines centrally.

With the advancement in networking technologies, these digital ledgers can be placed on multiple machines (PCs) and multiple persons can access and make the modifications to the record in the digital ledger. This led to the birth of distributed digital ledgers. However, a central server is present to approve any transaction added to the database (distributed digital ledger). This advancement did not stop here. Considering different applications and advancements in underlying network architecture and technology, a new paradigm shift occurred which advocates for the use of distributed digital ledgers in a completely decentralized manner. Thus, giving the birth to Distributed Ledger Technology (DLT). One major feature of DLT is that there is no need of clearing house to validate the transactions, as transactions are validated as soon as they are entered into the ledger. Table 1.1 shows the evolution of ledger technology.

1.1.1 Classification of Distributed Ledger Technology

In terms of identity of nodes, DLT can be classified as permissioned and permissionless. In permissioned DLTs, the identity of nodes needs to be known, while in permissionless DLTs, the identity does not need to be known. In terms of who can read the data of the ledgers, DLTs can be classified into public and private DLTs. In

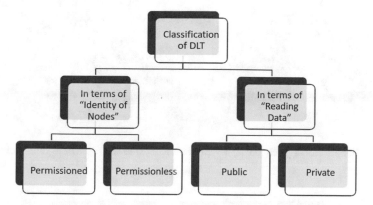

Fig. 1.1 Classification of distributed ledger technology with respect to identity of nodes and who can read the data of the ledgers

public DLTs, anyone can read the data, while in private DLTs, only approved node can read the data over the ledger. Figure 1.1 shows this classification.

1.1.2 Blockchain

In DLT, records and transactions are maintained in a distributed decentralized manner using Peer-2-Peer (P2P) networking technologies. The data management and organization in DLT can be done through various ways. For instance, data can be managed in the form of linear linked list of blocks or it can be managed using Directed Acyclic Graph (DAG) or tree-like data structures. If the data is managed in the form of linear linked list of blocks, then this is known as "Blockchain". One unique feature of blockchain is that it completely eliminates the role of trusted third party involvement in the maintainability of blockchain network.

> "Blockchain" can be defined as a data structure that is read only and data cannot be modified once it is entered into the blockchain and new data can only be appended at the end of blockchain, making blockchain highly immutable!

Blockchain technology operates over P2P networking. P2P is a different networking paradigm than client-server model of communication. This is one advantage of blockchain that it relies on P2P, which means no central entity is required to manage the network. Blockchain does not essentially be chain of blocks but it can be DAGs as well.

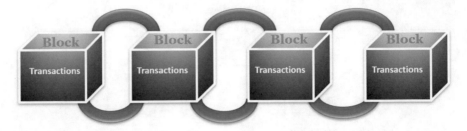

Fig. 1.2 Multiple blocks (each having several transactions) linked together to form a blockchain network

Blockchain systems can also be considered as trustless distributed networks.

In a blockchain network, a "block" is termed as a basic component in which transactions are assembled. Figure 1.2 shows a blockchain network in which multiple blocks are linked together (each block contains numerous transactions). These transactions are assembled in a block using cryptographic functions so that they cannot be tampered. Then, each of these blocks is linked together to form a blockchain. This linking can be performed in various ways. One simple way is to link these blocks in a linear order. However, there may be issues such as scalability, accessing these blocks quickly, and in terms of security. In order to address these issues, other structures can be used such as the blocks can be organized into graphs or trees. Figure 1.3 shows the contents of a typical block in a blockchain network, and blocks are linked together to form a blockchain.

Blockchain can also be considered as "state machine" replicated over several nodes.

Blockchain system is monopoly resilient, i.e., every node in blockchain can take participation in the decision-making process, thus making blockchain system democratized.

Transactions in blockchain can store information such as tracking property ownership, digital currency (cryptocurrency), loans, records of anything such as death records, birth records, and land records, tracking information of goods ranging from any sort of supply chain. Blockchain is "append-only" system, maintained by entities that do not fully trustable. "Ordered transactions" are maintained in blockchain in the form of log. Blockchain technology is a subset of distributed ledger technology.

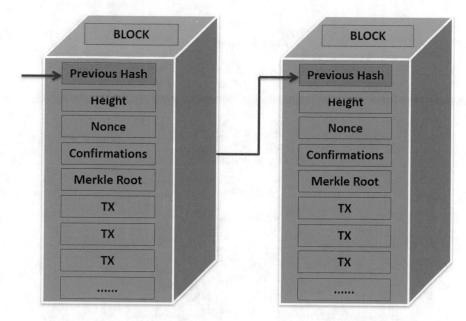

Fig. 1.3 Contents of a typical block in a blockchain network in which blocks are linked together with reference of hash value of the previous block

Figure 1.4 shows this relationship of blockchain technology with distributed ledger technology.

> In blockchain, transactions should contain some "value". For instance, tokens, cryptocurrency, commodities, and agreements for sharing assets can contain some value and transfer among the parties.

1.1.3 Directed Acyclic Graph (DAG)

In Directed Acyclic Graph (DAG) based DLTs, a DAG is formed to connect different transactions by a reference relationship. Example of such DAG based DLTs are Byteball, IOTA, and Nano. There are distributed ledgers which even do not follow block or DAG-based structures. They have their own unique data structures. Corda and Radix are examples of such DLTs. Transactions are directly stored in DAG by using the graph of transaction, instead of assembling these transactions in the form of block in blockchain. The processing speed of DAG seems quicker than blockchain as transaction do not need to assemble in blocks (as done in case of blockchain).

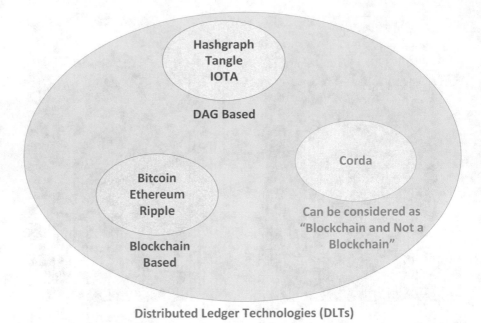

Fig. 1.4 Relationship between distributed ledger technology and blockchain. There can be DAG-based distributed ledgers, blockchain-based distributed ledgers, and the hybrid ones

Blockchain and DLT are used interchangeably and can be considered as a synonym, however, there are other ledger technologies which are based on DAG, etc.

1.2   Features of Blockchain Systems

Blockchain has some unique features that distinguish it from traditional database systems. Below these features are discussed in more detail.

1.2.1   Decentralization

In a blockchain system, there will be no central entity or intermediary to control and validate transactions. The data controlling capability is in the hands of users. This feature is known as decentralization.

1.2.2 Transparency

Blockchain systems (public blockchain in particular) are highly transparent. Anyone can track the transaction history and track transactions, thus making the blockchain system highly transparent.

1.2.3 Immutability

Once transactions have been added to the blockchain and validated by the participating nodes that transaction cannot be changed or tampered. This feature is known as immutability.

1.2.4 Availability

Due to the distributed and decentralized nature of blockchain, the ledger itself is available to nodes, thus making the system highly available as compared to centralized systems (with single point of failure).

1.2.5 Pseudonymity

Nodes in the blockchain system use pseudonymity, i.e., an identity that is partially revealed. Thus, making blockchain systems privacy aware.

1.2.6 Security

Security is an integral part of blockchain system. Strong public/private keys, hashing algorithms, digital signatures, and encryption techniques are used to secure the blockchain.

1.2.7 Non-Repudiation

Once a transaction has been added and validated in a blockchain, it cannot be disowned by the blockchain node. This makes blockchain system highly transparent.

1.2.8   Auditability

Auditability is another important feature provided by blockchain. It enables the user to trace any transaction within the ledger. In public blockchain, one can audit the whole ledger itself. However, in private blockchain, only authorized entities can perform this audit.

> One of the inherent features of blockchain is to provide service level agreement through smart contracts.

1.2.9   Data Tampering

One of the unique features of blockchain is the identification of data tampering. Blockchain stores the hash of the previous block in the current block. Then a process is carried out for every block in the blockchain in which current block is considered and its hash is generated. Therefore, any tampering in that data of any block in the blockchain results in the generation of a different hash value, which indicates that any tampering has been done in the chain. Thus, blockchain serves as a viable solution for the identification of data tampering. This can also be used to verify the assets owned by anyone as well as the transactions carried out by anyone.

> Blockchain in principle can be applied to exchange assets. These assets can be in any form ranging from digital currency to giving the rights to the blockchain users to only buy goods and services. Moreover, exchange can also be in the form of giving permission rights to let the blockchain user to participate in any activity such as voting. Blockchain can also be used in business automation. For instance, smart contracts can be integrated in the design of blockchain to support business automation. In supply chain, smart contracts can be integrated and transaction among different parties can be invoked once any particular event happens.

1.3 A Great Example for the Use of Blockchain Technology: Food Supply Chain

Food supply chain is one of the best applications where blockchain technology can be applied and the benefit can be seen for improving the system. As a customer, when we visit any retailers such as Tesco, Lidl, Aldi, or SuperValu, we have several choices of buying food products ranging from fresh fruits and vegetables grown locally to fruits and vegetables imported from other countries. There are also canned fruits and vegetables available. Besides this, we can also find ready-made meals made from the fruits and vegetables. We can also find these fruits and vegetables in frozen form. As a customer, our priority is to take that fruit and vegetable which is fresh, low prized, and with less contamination. We are sometimes also interested to see the origin of those fruits and vegetables. Few times we are also interested in checking the quality as well as any ingredients used in canned fruits and vegetables which were added during their processing in the food industry. We also check the expiry date and product details such as nutrition and energy level provided by that particular product. From religious perspective, we try to see that this particular food product is certified by which regulatory body, we see whether its Halal or not, and we also check the E-codes as well. Some people also check whether the food is suitable for vegetarians or not and some are interested in checking whether the food is organic or not. After looking at all these factors, we decide to buy any particular food product.

1.3.1 Traceability and Provenance Within Food Supply Chain

We, as a customer, are interested in all these aspects and these can only be possible if correct labeling is carried out. In addition to correct labeling, traceability and provenance of food are only possible if special care is taken at every step of the food supply chain from harvesting to delivery to the end customer. In European Union, standards are applied, and one cannot simply import every fruit or vegetable, instead special care need to be taken and standards are being implemented in the food supply chain. However, despite all these efforts, still we can find contaminated fruits and vegetables in the racks of retailers. And we often find that after the inspection of the regulatory body, Health and Safety Authority (HSA) in the case of Ireland. It is also possible that the complete lot has been recalled and the fruit or vegetable has been declared risky for customers. Though HSA ensures regular visits of its inspectors at various locations to identify such risky and contaminated food, but still few cases may be found about contaminated food, caused food poisoning to the customers.

It is also possible that the customers can also launch a complaint and report any such case of contaminated food. However, the problem at large is that once an item is identified as risky and contaminated, the whole lot/batch has to be removed from the whole country, from each and every retailer shop. Tracking all those lot/batches within a short period of time is not possible, so HSA and other authorities try to disseminate such news on national media channels such as newspaper, websites, social media sites, and also perform massive signage campaigns to alert the end users about that particular contaminated food. But still it takes few days to get rid from the complete risk-free food. It causes lot of financial loss for both the manufacturers and the retailers as well. This also has an adverse effect on the supplier as well and it deteriorates the supplier, manufacturer, and retailer trust. On top of it, identification of such a contaminated food item from the retailers shop also results in damage of the reputation of the retailer chain and such incidents, if happen frequently, decrease the trust of customers over the retailer chain and thus incur in a lot of financial loss.

1.3.2 Identification and Removal of Contaminated Food

Removing the complete lot/batch from the whole market is not a feasible solution and in order to avoid such future incidents, proper investigation needs to be carried out and it should be determined that food got contaminated at which level (retailer rack, retailer storage, supplier storage, manufacturer to supplier transport, supplier to retailer transport, or at the origin, i.e., the farmer or harvester?) And what was the reason for that contamination? Improper handling of food, exposing the food to sunlight, or lack of cold storage or any chemical reaction? These fine grained tracking and traceability along with provenance can only be possible with the integration of sensors for monitoring the food condition at every stage of the food supply chain and record the state of the food. Though this can be done using a traditional database but involvement of multiple parties within the food supply chain makes it challenging and difficult to maintain such a database. To illustrate the complexity, let's assume a database is maintained centrally by the retailer. However, since multiple parties are involved, therefore using this centralized database is not an optimal solution as who will control this database and it may also be considered as a single point of failure and also if harvester or farmer is not trusting the retailer after an incident then trusting on this data may not be possible.

1.3.3 Blockchain for Food Supply Chain

Blockchain can be used in food supply chain and due to its inherent features of distributed and decentralized nature, all the parties involved in food supply chain can trust on each other and also the data recorded on the blockchain will be immutable and tampering cannot be possible. Blockchain solution for food supply chain will ensure the provenance and traceability of the food products as well. This will ultimately help the inspectors (HSA in case of Ireland) to track any food item and ensure that full compliance has been made at every step of the food supply chain.

1.4 Summary

In this chapter, we discussed the background of ledgers and highlighted how distributed ledger technology evolved with the passage of time. We then focused on blockchain (a distributed ledger) and mentioned features of a blockchain system. To demonstrate the effectiveness of blockchain technology, we presented a food supply chain example and use of blockchain in it. In the next chapter, we will be discussing how blockchain technology is superior to database management systems.

1.5 Further Reading

The goal of this section is to highlight some related work and if the reader is interested to explore further these topics, the following references may be very useful.

1.5.1 General Blockchain History and Background

To further explore general blockchain history and background, [18] and the references therein is a wonderful resource.

1.5.2 Food Supply Chain and Blockchain

Further reading about food supply chain and blockchain can be found in these references: Internet of Things and blockchain-based food supply chain is discussed in [24, 62, 69, 70, 84]. Other important references on food supply chain are [1, 16, 25, 28, 35, 47, 59, 65, 71]. Blockchain, food supply chain economics, and its adoption in China can be found in [54–56]. A discussion on supply chain of things can be found in [96]. Grain supply management and its safety using blockchain is discussed in [103]. Using smart contract for product traceability can be found in [90]. US Beef Cattle supply chain is discussed in [30]. A discussion on trade supply chain can be found in [45]. Blockchain in the context of smart industry can be found in [20]. Scalability issue of blockchain enabled supply chain can be found in [61].

Problems

1.1 What are the alternatives to blockchain?

1.2 The current financial systems and Internet transactions model is working fine. Describe why there is a need of blockchain technology without a centralized authority?

1.3 How assets were managed before the digital computer systems?

1.4 How directed acyclic directed graph differs from blockchain?

1.5 What is the difference between blockchain technology and distributed ledger technology?

1.6 Explain how blockchain is a disruptive technology?

1.7 Describe five features of a blockchain network.

1.8 How Fiat currency is different from cryptocurrency?

1.9 Explain in your own words the need of blockchain.

1.10 What are blockchain-based distributed ledgers and what are directed acyclic graph-based distributed ledgers?

1.11 Provide a comparison between banking model and Bitcoin.

1.12 Explain how blockchain is evolved over time.

1.13 Describe a scenario where you can apply blockchain.

1.14 Bitcoin and Ethereum are very high prices in the market. These systems require very reliable system. In this context, comment on the reliability of blockchain system.

1.15 Explain in your own words how blockchain is a viable solution for food supply chain?

1.16 Explain how blockchain can support food safety?

1.17 How blockchain can be considered as Internet of Transactions?

1.18 How data tampering is difficult in blockchain systems?

1.19 How blockchain systems are publicly auditable?

Chapter 2
Blockchain Technology and Database Management System

2.1 Distributed Ledger Technology and Database Management System

DLT is basically a database, however, this database is not like a traditional Database Management System (DBMS). In conventional DBMS, a central entity is responsible for keeping the records up to date. This central entity is also responsible for validating any new transactions that need to be recorded in the database. Similarly, if any entry needs to be renewed from the database, nodes need to get it done via the central entity. Moreover, the nodes do not keep the whole copy of the database. On top of it, due to this centralized nature, DBMS is prone to single point of failure.

When we think about DLT as a database, it replicates the same features of database in it, however, it differs in few aspects (cf. Table 2.1). For instance, DLT maintains a database among the nodes (geographically dispersed) by replicating the whole database to the node itself. This feature makes DLT very much error and failure prone compared to the traditional databases. Moreover, each node will have a global view of the whole database. This feature also leads DLT to attack resilient, as the attacker will now require tampering a huge number of nodes compromised to make any change in the ledger copy. Additionally, it makes DLT more robust against unavailability, compared to DBMS.

One unique aspect of DLT compared with traditional DBMS systems is the capability of reaching consensus among the participating nodes. This feature of consensus is not present in DBMS. In a traditional database, when an entry or transaction is required to be included in the database (DB), not all the nodes need to reach to a consensus, instead the central DB server or entity just need to approve this transaction by checking the credentials of the node. It needs a layer of trust which is essential between the participating nodes and the DB server. This is not the case with DLT, where a trust layer is not present among the nodes (cf. Sects. 3.16.1 and 4.1 for more details on trust). Basically, a node who wants to include a transaction entry in the ledger needs to propagate this to the DLT participating nodes. These participating

© Springer Nature Switzerland AG 2021
M. H. Rehmani, *Blockchain Systems and Communication Networks: From Concepts to Implementation*, Textbooks in Telecommunication Engineering,
https://doi.org/10.1007/978-3-030-71788-9_2

Table 2.1 Unique features of Databases and DLT

Feature	DBMS	DLT
Common Records	Yes	Yes
Central Entity	Yes	No
Consensus Management	Through central entity	Through mining nodes
Duplication of Data	Few nodes or central entity	Kept by every node
Global View	Yes	Yes
Distributed	Yes	Yes

nodes then validate this transaction by using a consensus algorithm (cf. Chap. 4). Once the transaction is validated, it will then be appended to the distributed ledger. In this manner, the DL is updated and synchronized among all the nodes in the system.

Blockchain can be considered as a database but with unique features as mentioned in Sect. 1.2. Therefore, virtually it can be applied to any application area in order to replace databases, however, the feasibility of adopted blockchain technology needs to be evaluated before deciding any potential use of this technology to a particular application.

Oracle, MySQL, and other database systems are used in applications like finance, asset management, insurance, and banking system. However, with blockchain's inherent features, these systems can become more transparent, incur low cost, and even reduce the level of human intervention by making system more automatic, and thus existing infrastructure can bring economic benefits (see reports available on Internet by famous financial companies around the globe).

> Traditional database systems are designed to handle with simple crash failure; however, blockchain is designed to handle even more severe hostile environment, i.e., Byzantine environment.

Differentiating database systems with distributed ledgers can also be understood from the perspective of state of the ledger. In distributed ledger system (blockchain), the next state of the ledger is achieved with the help of consensus algorithm. In public blockchain system, this new state of the ledger is achieved by reaching consensus among all participating blockchain nodes, while in consortium or private blockchain, ledger blockchain nodes are elected and responsible by proposing the new state of the ledger. This new blockchain state is then communicated to all the nodes in the network.

> In traditional database systems, the order of transactions does not matter a lot, while order of transactions is of utmost importance in blockchain systems.

We can think blockchain as an alternative way of storing data. In comparison with Relational Database Management System (RDBMS), blockchain ensures that the adding, deleting, and updating of records is not within the hands of single entity, instead this addition, deletion, and updating of records can be carried out in a trustless environment. Such an approach removes the reliance of blockchain system to one or few entities, thus decreasing the chances of data tampering and central point of failure. Another issue with blockchain system is associated with the blockchain governance model and update of software. Let's take an example of public blockchain, since nobody owns the public blockchain, thus updating the software also requires consensus at all blockchain levels.

Compared to traditional database, the fees for adding transaction in blockchain are high and changes dynamically. Thus, one may need to think about transaction fees when adding all the transactions to the blockchain.

In blockchain system, miners spend lot of energy to solve the puzzle for adding the block mechanism together with difficulty in solving the puzzle (that requires lot of energy consumption), lead the miner nods to behave correctly.

Imagine if YouTube operates on public blockchain then no (central YouTube) authority can control the data uploaded on YouTube. Thus, the censorship problem that we may find in some countries can be reduced to certain extent. However, removing such a central regulator (YouTube) admin may change the business model and it may be difficult to remove inappropriate content from YouTube.

When we think about replacing traditional database with blockchain technology, we need to understand that there are few implications as well. For instance, the transaction speed is slower in blockchain as compared to traditional database systems. Moreover, we need a trusted third party who manages a database, while in blockchain system, the role of this trusted third part is not present but this results in who will update the blockchain software?

2.2 When to Select Blockchain Over DBMS?

We need to answer two important questions which helps us to decide which technology to select between blockchain and DBMS.

- When to select blockchain over DBMS?
- When to select permissioned or permissionless blockchain?

Traditional database systems or spreadsheets generally store the information very well in a very structured way. Moreover, Structured Query Language (SQL) and other data mining algorithms are advanced so much that complex queries can be handled easily and desired information can be extracted from the databases quickly. However, in database, initial credential checking is required to access and retrieved the data. The same case applies when merging, sorting, modifying, or deleting data from the database. In database systems, database state is not shared among the participating database nodes. Additionally, in database systems, not every entry is secured and it is not designed to backtrack any previous transactions entered into the database. Contrary to this, blockchain systems provide the state of the ledger to be shared among all the participating nodes. Moreover, every single transaction is cryptographically secured. Therefore, when we decide to adopt whether blockchain or database system should be our choice, we need to consider whether we need to cryptographically secure each and every entry of the transaction or not and whether we need to share the state of the ledger among all the nodes or not. Another feature that helps us to decide which technology blockchain or traditional database is suitable for our scenario is the modification rights to the ledger. It is the unique feature of blockchain that all the nodes have written access to the ledger.

2.3   Blockchain and Database Maintenance

Blockchain and database maintenance is another criterion. We may consider when deciding about technology selection. In centralized database systems, maintenance is the responsibility of central entity and this central entity needs to be highly reliable and trusted entity. In distributed database systems, multiple designated nodes are responsible for database maintenance. In blockchain system, there is a lack of trusted entity and ledger need to be maintained in the presence of such nodes.

2.3.1   Ledger Maintenance in Public Blockchain

In public blockchain, ledger is maintained by all the nodes collectively. There needs to be an entity which resolve any conflict or makes protocols but in principle, any modification or maintenance will be the collective responsibility of all the participating blockchain nodes.

2.3.2   Ledger Maintenance in Consortium Blockchain

In consortium blockchain, a consortium of nodes is responsible for updating and maintaining the blockchain.

2.3.3   Ledger Maintenance in Private Blockchain

In private blockchain, only a single node is responsible for blockchain maintenance. It resembles the centralized system.

2.4 Database System, DLT, and Public Verifiability

Public verifiability means you want your system (ledger) to be verifiable by the public, i.e., public can verify the state of the system at any time to trace and track any transaction or ledger entry. Of course, if this feature is not required then public blockchain is not an option.

If we require a system (ledger) where this public verifiability is not required then we may consider the design choice among consortium and private blockchain and database systems. Of course, traditional database will be a good option when we require a central entity to control the state of the ledger or nodes need approval from this central database entity. As we discussed earlier in Sect. 2.1, this may also lead to single point of failure. On the contrary, consortium and private blockchain provide the feature of organizing the ledger using the underlying P2P architecture.

In traditional database systems, central server is responsible for managing the database. It has some clear advantages such as low cost (software), easy to manage, higher throughput (processing speed), widely adopted and tested since few decades, and changes can be seen quickly by the nodes as central entity process the updates quick and those changes are visible to the nodes.

In blockchain system, P2P underlying network architecture is running at the back to manage the system. Similarly, new design patterns (software) are required for blockchain-based systems. This not only increases the initial deployment cost but blockchain systems are also prone to low transaction processing speeds.

2.5 Comparison of Blockchain Systems and Traditional DBMS

There are some unique inherent features of blockchain system (see Sect. 1.2). We compare these features of blockchain systems with DBMS.

1. **Hashing**: In traditional database system, hashing is not applied to each data entry in the database, however, in blockchain system, hashing plays a critical role and applies to different stages (see Sect. 3.26 for more detail).
2. **Public/Private Key**: Public/private encryption is generally not applied in database systems; however, this is an essential requirement in blockchain systems.
3. **Sequence of Records**: In database systems, records and transactions are as such not maintained sequentially, as in blockchain systems. It is a strict requirement in blockchain system that transaction should be placed sequentially and linked together.
4. **Consensus**: Consensus is not required in database systems to maintain and update the state of the ledger. However, in blockchain systems, it is an essential requirement.

Table 2.2 Large-scale databases and DLT

Name of framework	Database/DLT	Architecture
Hadoop Distributed File System (HDFS)	Database	Master-slave
Hyperledger Fabric	DLT	P2P
Hyperledger Sawtooth	DLT	P2P
OpenChain	DLT	Client-server
MultiChain	DLT	P2P
HBase	Database	Master-slave
Cassandra	Database	P2P
MongoDB	Database	P2P

5. **Immutability**: Blockchain records are immutable, i.e., no previous entry or transaction recorded in the ledger can be modified. This feature is not present as such in database systems.
6. **Ledger Dissemination**: In traditional database systems, once any update is made in the ledger, the state of the ledger is not required to disseminate among the participating nodes. However, this dissemination is essential in blockchain systems.

2.6   Large-Scale Distributed Database Systems and Blockchain

Large-scale distributed database systems, such as Apache Cassandra and Hadoop Distributed File System (HDFS), are available and have been utilized extensively in real implementations. The question is which one is more powerful and better over the other: distributed database or blockchain? To answer this question, we need to see the underlying architecture of database systems and blockchain, both need to be same. To illustrate this, let us say we compare blockchain system which is operated on P2P network and if we compare it with a client-server architecture-based database system such as HDFS or HBase then this comparison does not look fair. Table 2.2 compares large-scale DBMS and DLT frameworks. Second important aspect when comparing database with blockchain is the feature of keeping the copy of record by the node. This is the primary feature of blockchain that every node will keep the copy of the ledger, while in database, a different terminology is adopted which is replication strategy. So, when comparing database system with the blockchain system, the replication strategy also needs to be considered. When comparing database with blockchain system, factor such as how frequently user insert or update data and how quickly it gets the data also need to be considered.

2.7 Trust and Public Availability of Blockchain

Trust and public availability of blockchain systems is one such unique feature that distinguishes it heavily from traditional database systems. Blockchain, in essence, remove the trusted third party from the middle among the transacting parties and build trust publicly in un-trusted and unknown environment. Moreover, the public availability and transparency makes blockchain system more popular among their traditional counterparts, i.e., database systems.

2.8 How Blockchain Is Different from Distributed Data Storage?

Data storage technologies have also been evolved over time. Initially, data storage was carried out in a centralized manner, but due to advancement in networking technologies, processing power, and hardware, data storage is managed in a distributed manner. Means, multiple data storage servers hold the same copy of the storage and they are responsible to delete, modify, or add records to the data server. This data storage server then synchronizes with other data storage servers so that the same copy of the data being replicated to all storage servers. Blockchain system is different from distributed data storage in a sense that it eliminates the need of a centralized entity. Moreover, blockchain systems differ from distributed data storage with all its unique features (cf. Sect. 1.2 for more details).

2.9 Summary

In this chapter, we discussed in detail the differences between blockchain technology and database management systems. We also discussed the criteria for the selection of blockchain. Furthermore, a comparison of blockchain systems and database management system is provided. We also discussed how blockchain differs with database management systems from the perspective of trust and public availability. Finally, the chapter concludes by highlighting the differences between blockchain and distributed data storage.

2.10 Further Reading

Further reading about blockchain and databases can be found in [19]. The article [95] is a nice resource for the readers who are interested to understand further in detail when to select the blockchain.

Problems

2.1 Suppose there are $N = 20$ consensus nodes in a blockchain network. This blockchain network can reach to consensus - $\lfloor \frac{N-1}{5} \rfloor to \lfloor \frac{N-1}{2} \rfloor$ toleration level - in the presence of Byzantine nodes. How many Byzantine nodes (faulty nodes) this blockchain network can tolerate?

2.2 You want to work on Ethereum blockchain. How you find the size of Ethereum blockchain?

2.3 Blockchain systems support different transaction handling capacity. How to find the transaction handling capacity of blockchain?

2.4 How many copies of the distributed ledger should be disseminated in a DLT?

2.5 Why there is a need for blockchain system when we already have DBMS?

2.6 What are the unique features of DLT and DBMS?

2.7 In which cases we prefer DLT over DBMS?

2.8 How blockchain and database systems are managed?

2.9 Is public verifiability is possible in DBMS?

2.10 Provide a comparison between blockchain system and traditional DBMS.

2.11 How large scale database systems differ from blockchain systems?

Chapter 3
Blockchain Fundamentals and Working Principles

In the starting of Internet era, the focus was on making resources and computing available in a centralized manner. This led to the most popular computer design architecture, i.e., client/server model. Through this model, applications were designed, and memory, storage, and computing were all following the same architecture. However, with more computing power available at the consumer end and requirements of customers, this client/server architecture was being replaced with peer-to-peer architecture and now with the advancement in fog/edge computing, and cloud computing, computing, storage, and other resources are getting distributed and decentralized. This paradigm shift cannot be possible without underlying technologies and blockchain is one such technology that is helping to achieve this realization completely.

3.1 Blockchain Network

Blockchain networks can be classified as public, private, or consortium. It can also be classified as permissioned blockchain and permissionless blockchain. The permissioned blockchain can be further classified into private blockchain and consortium blockchain. The permissionless blockchain is also referred to as public blockchain. Figure 3.1 shows the classification of blockchain network as public blockchain, private blockchain, and consortium blockchain.

In permissioned blockchain, the access granted to the blockchain nodes is limited. In simple words, we can say that not everyone can get access to the permissioned blockchain over the Internet. This access limitation over the Internet can be further restricted and controlled. If a set of companies make a consortium, then access to this blockchain is subject to the approval of this consortium; such type of blockchain is called as consortium blockchain. Similarly, if access to the blockchain is completely invisible over the Internet and only private node(s) grant access to the blockchain, then such type of blockchain is referred to as private blockchain.

© Springer Nature Switzerland AG 2021
M. H. Rehmani, *Blockchain Systems and Communication Networks: From Concepts to Implementation*, Textbooks in Telecommunication Engineering,
https://doi.org/10.1007/978-3-030-71788-9_3

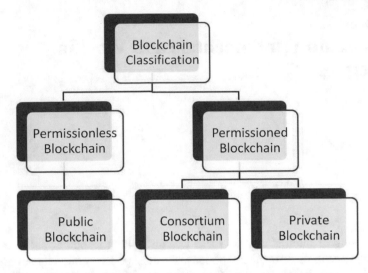

Fig. 3.1 Classification of blockchain network as permissioned and permissionless blockchain. It is further classified into public blockchain, private blockchain, and consortium blockchain

3.1.1 Public Blockchain Network—Permissionless

A public blockchain network—permissionless blockchain—is the one in which any blockchain node can join or leave at any time and participate in mining process. This type of blockchain network is completely decentralized. However, this is not true for the case of permissioned (consortium) blockchain. An important aspect that distinguishes permissionless blockchain over database system is its inherent features (See Sect. 1.2) that still make permissionless blockchain superior in few cases over the database systems.

3.1.2 Private Blockchain Network—Permissioned

A private blockchain network is the one in which nodes can join or leave the network with the permission from the central entity. Such type of blockchain network is completely centralized.

In permissioned blockchain, miners are selected a priori.

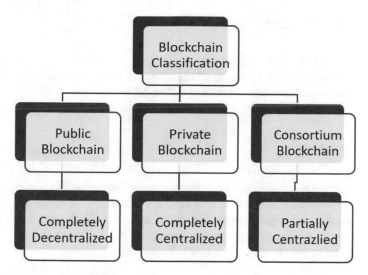

Fig. 3.2 A more refined classification of blockchain network from the perspective of centralization and decentralization

3.1.3 Consortium Blockchain Network—Permissioned

Consortium blockchain network is neither completely centralized nor completely decentralized. Nodes can join or leave the blockchain network with the permission from the consortium of nodes (who manages the blockchain network). In permissioned blockchain, nodes have limited role as compared to permissionless blockchain. Not all the nodes can validate the transactions. Instead, a set of validator nodes can be selected for validation process. It is also possible that a set of consortium nodes being elected publicly for the validation of transactions. This will have a high impact on the Tx handling capacity of blockchain. Since only a set of nodes will be responsible for validating transactions, therefore, transaction addition speed will be faster. Moreover, since only few nodes will be involved in the mining process, this process can be done quickly.

Figure 3.2 shows the classification of blockchain network from the perspective of centralization and decentralization.

3.2 General Issues with Public Blockchain

Public blockchain has few general issues such as limited transaction speed, scalability, pseudonymity, limited block size, and energy consumption during the mining process.

3.2.1 Limited Transactions

In a public blockchain, the transaction processing capability is still limited (see Table 3.8). Bitcoin blockchain can handle less than 10 transactions per second (Tps), while Ethereum blockchain can handle less than 50 Tps.

3.2.2 Scalability

Scalability (in terms of number of blockchain participating nodes) is another issue in public blockchain. When public blockchain systems are designed for different communication technologies such as Internet of Things (IoT), the amount of participating blockchain nodes will increase exponentially. This will have an adverse effect on mining time as well as size of the blockchain.

3.2.3 Pseudonymity

In public blockchain systems, complete anonymity is not suggested, for instance in Bitcoin. The primary reason for doing this is to deal with blockchain specific attacks such as double spending and Sybil attacks. This partial visibility of identity may also trigger other privacy attacks and malicious nodes may exploit this vulnerability.

3.2.4 Block Size

Public blockchain to date has limited block size. For instance, Bitcoin has 1 MB Block size.

3.2.5 Energy Consumption

Energy consumption is another important aspect of public blockchain. Due to the reliance of public blockchain system (such as Bitcoin) on PoW cryptographic puzzle, a huge amount of energy is consumed.

3.3 Underlying Network for Peer Discovery and Topology Maintenance in Blockchain

Peer discovery and topology maintenance is an essential aspect to run the blockchain network. In Bitcoin blockchain network, peer discovery, and topology maintenance are done through volunteer DNS servers, while in Ethereum blockchain, Distributed Hash Table (DHT) based protocols such as Kademlia are used for peer discovery and topology maintenance.

3.4 Broadcasting in Blockchain Network

Broadcasting means dissemination of information to all the nodes in the network. Broadcasting is useful in realizing many useful networking tasks. For instance, broadcasting is used for neighbor discovery, broadcasting is used in routing protocols, and broadcasting is also used to share control information among all the nodes in the network. In the context of blockchain network, broadcasting is also used to share the ledger copy among all the nodes, which means sharing transactions and blocks. Looking at the life cycle of blockchain transactions, when a node makes a transaction, it broadcasts this transaction to all the nodes. These transactions are then stored in Mempool. The miner nodes then assemble few transactions to make a block. Once a block is successfully mined; it is then broadcast to the whole blockchain network. Then the nodes locally update this ledger copy. In this manner, broadcasting helps nodes in the blockchain network to update their copy of the blockchain. Since the blockchain network operates over the P2P network, therefore, underlying P2P protocols are used to broadcast this information. P2P protocols use gossip-based protocols for such dissemination. In Ethereum, DevP2P Wire protocol is used for such purpose.

3.5 Users/Nodes in a Blockchain Network

Blockchain users/nodes can make transactions and transfer assets through their application. Each blockchain user/node has to generate public/private key pair for authentication purpose. In order to become the part of the blockchain network, a new blockchain user/node has to establish a minimum number of peer connection (direct connections) with other nodes in the blockchain network. Then and only then the newly joined nodes will be considered as part of the blockchain network. Table 3.1 shows the types of blockchain nodes and their involvement in different aspects within the blockchain system.

Table 3.1 Types of blockchain nodes and their involvement in different aspects within the blockchain system

Copy of ledger	Full node	Lightweight node	Miner node
	Keeps complete copy locally	Keeps partial copy locally	Keeps complete copy locally
Transaction generation	Yes	Yes	–
Consensus participation	–	–	Yes
Routing for message dissemination and verification	Yes	Yes	Yes
Transaction validation	–	–	Yes

There are three types of blockchain nodes:

- Full blockchain nodes,
- Lightweight blockchain nodes, and
- The miner nodes or consensus nodes or miners.

3.5.1   Full Blockchain Nodes

Full blockchain nodes are those nodes which keep the full copy of the blockchain ledger locally. They also participate in transaction verification without reference to any node externally. Full blockchain nodes also participate in routing.

3.5.2   Lightweight Blockchain Nodes

Lightweight blockchain nodes do not carry the complete copy of the ledger, instead they only keep the header of each block and then reference to the external nodes when complete information is required. Lightweight blockchain nodes also participate in routing.

3.5.3   Miner Nodes

Miner nodes are those nodes which carry the consensus mechanism and responsible to publish the block in the blockchain. In this manner, miner nodes are the most powerful nodes in the blockchain network, as they can change the state of the blockchain network.

A node in a blockchain, which participates in consensus mechanism, i.e., miner node can only create and add a block to the blockchain.

3.6 Blockchain Nodes as Leaders and Validators

Leader nodes are those blockchain nodes that create blocks. Validator nodes are those blockchain nodes which validate the blocks. Depending upon the type of blockchain, the role of leader nodes and validator nodes can be defined. In some blockchain, any node can be served as a leader node, i.e., create blocks and any node can be served as a validator node. A typical example of such a case is public blockchain (permissionless blockchain). In other blockchains, the role of leader node and validator node can be separated. For instance, in permissioned blockchain, leader and validator node's role can be separated.

Validator nodes can also be called as miner nodes. In contrast, the leader node can be elected and serves during a specific time. This specific time varies from blockchain implementations.

3.7 Blockchain Nodes as Sender and Receiver

When a blockchain node makes a transaction, this node can be referred to as a sender node. When creating the transaction, this sender node has to specify the name of destination node, i.e., receiver node. Both sender node and receiver node can be identified using addresses. It may be possible that a blockchain node (sender node) issues a transaction to multiple blockchain nodes (receiver nodes). It is worth noting here that the transaction originator (sender node) has to sign the transaction with its private key, which is an essential step in blockchain verification and validation process. Without signing, transactions may not be processed further. Previously (see Sect. 3.5), we discussed three types of blockchain nodes: full nodes, lightweight nodes, and miner nodes. In terms of dealing with transactions, we can have sender nodes, receiver nodes, leader nodes, and validator nodes. Table 3.1 mention the roles of nodes and their involvement in transactions handling.

3.8   Layers in Blockchain

Figure 3.3 shows layer in blockchain system. We can divide the blockchain system into six layers from bottom to top: hardware layer, data organization and topology layer, network and operating system layer, consensus layer, virtualization and smart contract layer, and application layer. Below we discuss each layer and its purpose in more detail.

3.8.1   Application Layer

This is the topmost layer of blockchain. It directly interacts with the user. In this layer, different applications run over the client and the client uses the applications to interact with the blockchain system. It contains applications such as cryptocurrency, DApps, wallets, and other domain specific applications including 5G and IoT-based applications. Wallets are required to manage users account, and this includes management of tokens (cryptocurrency) and generating and receiving transactions.

> In Hyperledger Fabric, smart contract runs on Docker (in a container).

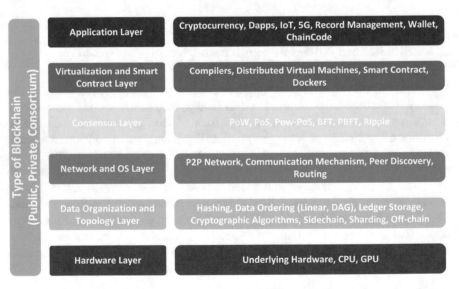

Fig. 3.3 Layers in blockchain. We can divide the blockchain system into six layers from bottom to top: hardware layer, data organization and topology layer, network and operating system layer, consensus layer, virtualization and smart contract layer, and application layer

3.8.2 Virtualization and Smart Contract Layer

This layer is responsible for virtualization and smart contract related execution. All the compilers which execute Ethereum Byte code, and run smart contracts execute over this layer. This layer also interacts with the user machine and responsible to compile blockchain code. It provides environments such as Ethereum Virtual Machine (EVM) and Java Virtual Machine (JVM). When a user installs blockchain client (e.g., wallet) on his/her machine, the compiler (virtual machine) also installs on it, which is responsible to compile all the code generated over the wallet. When a user wants to work on Ethereum blockchain, Ethereum virtual machine is responsible to compile smart contract code (e.g., Ethereum Byte code). Ethereum smart contracts are written in an open source programming language named as Solidity. Similarly, when the user wants to run Hyperledger Fabric, it can be executed on Java. Node.js or even in Go programming languages. In Hyperledger Fabric, smart contract is knowing as chaincode and runs inside Docker.

3.8.3 Consensus Layer

Consensus layer is responsible to manage and reach consensus in a P2P network. This layer operates and interacts at network level in the blockchain network. It dictates which consensus protocol needs to be executed and how to follow the rules to achieve consensus. Different types of consensus protocols are available so depending upon the adopted blockchain system, a particular consensus algorithm will be selected and executed by this layer. In blockchain systems which provides blockchain as a service (BaaS), a variety of consensus protocols are available as Plug-And-Play (PnP). For instance, Microsoft Azure supports several blockchain platforms such as Ethereum, Corda, Hyperledger Fabric, and their corresponding consensus protocol implementations. Details and working of consensus protocols can be found in Chap. 4.

3.8.4 Network and OS Layer

This layer is the core layer responsible for managing the underlying network services and operations over a blockchain network. In this layer, communication mechanism, peer discovery, routing, and peer-to-peer network are managed. Peer-to-peer network is the core network over which blockchain network runs.

Table 3.2 Storage and data structure models for blockchain

Blockchain	Data model
Hyperledger Fabric	Bucket-Merkle Tree (to store indices)
Ethereum	LevelDB (to store states)
Hyperleger Fabric	CouchDB (to store states)
Parity	Patricia Merkle (to store key value)
Ethereum	Patricia Merkle (to store key value)
IRI	RocksDB (to store snapshot)
IOTA	DAG

3.8.5 Data Organization and Topology Layer

This layer is responsible for data organization and topology management. It includes
tasks such as hashing, data storage (ledger storage), cryptographic algorithms, data
ordering (linear, DAG), side chain, sharding, and off-chain related issues. It also
includes issues such as transaction models, and Merkle Tree management. There are
two types of transaction models: (i) Unspent Transaction Output (UTXO) and (ii)
account. More details about transaction models are presented in Sect. 3.24. Table 3.2
shows storage and data structure models for blockchain.

3.8.6 Hardware Layer

This layer deals with the underlying hardware of the blockchain nodes. The performance of blockchain nodes depends upon the architecture used, i.e., whether its CPU
or GPU, etc.

3.9 General Working Sequence of Blockchain

Figure 3.4 shows contents of block in a blockchain network showing specifically
block header and genesis block. Below we discuss the general working sequence of
blockchain.

1. Blockchain user(s) creates account(s).
2. Blockchain user creates transaction (Tx).
3. Transactions are signed.
4. Transaction is broadcast to the validating nodes.
5. Transaction is validated.
6. Transactions are gathered in a pool.

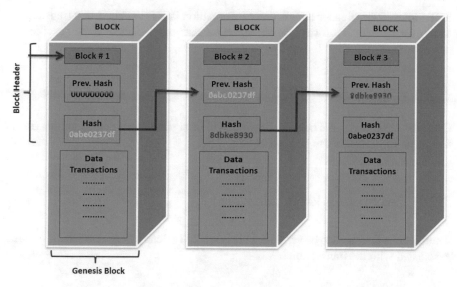

Fig. 3.4 Contents of block in a blockchain network showing specifically block header and genesis block

7. Miner node(s) gather multiple Txs and create block.
8. Miner node(s) perform mining.
9. Block validation is carried out.
10. Successful miner node adds block to the blockchain.
11. Added block information is broadcast to the blockchain network.
12. Blockchain node adds the broadcast block to their local copy of blockchain—Block confirmation.
13. Block becomes part of the global blockchain—Block published.

3.9.1 *Transaction*

In blockchain, transactions can be of any form. It can be in the form of cryptocurrency, digital asset, land record, buyer or seller information, and any other information. Depending upon the application area and deployment of blockchain, transactions contain different information. For instance, when deploying blockchain in telecommunication for spectrum auction, it may contain information about the spectrum and lease (See Chap. 6 for more information). Transactions may also contain a set of computing instructions to perform. A user can create a single transaction or multiple transactions. These transactions may require different information from the blockchain user at a time of transaction creation and may vary from blockchain implementations. General information such as the asset to transfer needs to be specified in the transaction. Sender and receiver information is also required in a transaction. Moreover, redeemed information about the transaction also needs to be specified.

3.9.2 Transaction Signing

Once the transactions are created, they need to be signed. More details on this transaction signing process can be seen in Sect. 3.23.

3.9.3 Transaction Verification

Transactions also need to be verified. This is typically done by checking the signature associated with the transaction.

3.9.4 Transaction Broadcast

Transactions are broadcast in the blockchain network. Validating nodes validate the transactions and add them to form a block. The speed of this broadcast has an impact on the overall performance of the blockchain system.

3.9.5 Transaction/Block Validation

Transaction validation step occurs at different stages. In the first stage, when a blockchain user creates a transaction and broadcast it to the blockchain network, the transactions are validated. Later, another time, the process of transaction validation occurs. This time, transaction (in the form of block) is validated. We may say that this is block validation. Block validation is the responsibility of miner nods. These miner nodes, using consensus algorithm, validate the block.

3.9.6 Block Confirmation

Block confirmation is carried out once the block is successfully mined. Once the block is confirmed and published in a blockchain, it can no longer be changed. If any details in the block must be modified, then a new block needs to be added to correct such information in the blockchain.

3.10 Composition of a Block

A block is composed of several things. Figure 3.5 shows the components of a block, which we describe below:

3.10.1 Hash Pointer

Hash pointer is responsible for pointing out the previous hash so that the order of blocks in the chain is maintained. This same hash pointer can also be used to verify the integrity of the blockchain. This hash pointer of the previous block is created by concatenating all the fields of the previous block and represents by hash function.

3.10.2 Merkle Tree

The reason for using Merkle Tree is to verify the set of all transactions quickly within a block. Merkle Tree is organized in the form of tree.

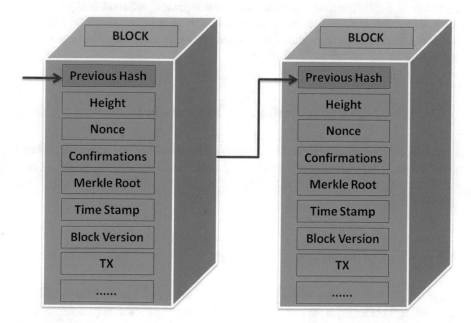

Fig. 3.5 Detailed contents of block in a blockchain showing time stamp, nonce, block header, body, Merkle root, and hash pointer

For each transaction, a separate hash value is calculated. Then, the two hash values are combined to generate a new hash going upward toward the root. Then at the root, all the hash values are combined to create a new hash value. This hash value is then used for integrity and verification purpose. Now, if this hash value has been modified, it means, any transaction within the block has been modified.

Merkle Tree is a binary tree used to represent transactions in a block. Merkle root is known as the root of the Merkle Tree wherein each leaf is labeled as hash code of the transaction. The hash code of two children is combined together as label of the non-leaf node.

> Blockchain is secured at different levels. First, hashing is applied, and a root hash is generated. For instance, Bucket Hash Tree is used in Hyperledger V0.6 while Patricia Merkle Tree is used in Ethereum.

3.11 Blockchain Governance System: Who Owns Blockchain?

In terms of deployment and involvement of companies, different companies can operate in a consortium blockchain. However, if there is only a single organization involved, then it can be considered as a private blockchain. Corda and Hyperledger are examples of consortium blockchain.

3.12 Who Make Modifications in Blockchain?

In terms of ledger modification, blockchain can be classified as permissionless and permissioned. In permissionless blockchain, anyone can modify the ledger. In permissioned blockchain, restricted number of nodes can make modification in the ledger.

3.13 Confidentiality in Blockchain

Public blockchain systems do not provide confidentiality except the blockchain user and their actual identities are not disclosed to the general public using pseudonymity. It means, in public blockchain systems, the complete ledger along with transaction history is available to the public except the real identity of blockchain user. On the other hand, consortium and private blockchain provide more confidentiality to the blockchain users, i.e., public cannot trace and track and even access the blockchain transactions. However, blockchain nodes need more trust in such an environment, and the identity of the nodes should be known within the network. In simple words, we may say that nodes within the consortium and private blockchain system known each other and publicly they are completely hidden.

3.14 Blockchain Platforms

Table 3.3 shows the application of blockchain and different blockchain systems which supports such applications.

3.14.1 Availability of Blockchain Platforms

Github is one of the primary sources where one can find several open source blockchain code-bases. However, before selecting any one of them, substantial knowledge is required for the selection among these code-bases. Hereafter, we discuss few of the existing blockchain platforms.

Table 3.3 Application of blockchain and different blockchain systems which supports these applications

Application of blockchain	Blockchain systems
Asset Management and Tracking	Corda
Asset Management and Tracking	Bigchain DB
Asset Management and Tracking	Multichain
Own assets in terms of Token	Stellar
Own assets in terms of Token	IOTA
Cryptocurrency	Bitcoin
Cryptocurrency	Litecoin
Smart Contract-based new Business Logic	Parity
Smart Contract-based new Business Logic	Ethereum
Smart Contract-based new Business Logic	Quorum

3.14.2 Blockchain Platform Suitable only for Cryptocurrency

Several blockchain platforms are available online that provide the basis to build our own cryptocurrency on top of it or to get insights about the working of these existing blockchain networks.

3.14.3 Blockchain Platform that Supports Smart Contracts (Business Logic)

Business logic can also be integrated to the blockchain with the help of smart contracts.

- Bitcoin is a blockchain that supports transactions only and it operates in permissionless environment.
- Chain Core is a blockchain that supports transactions only, however, it is basically a consortium blockchain.
- Ethereum is a permissionless blockchain that supports smart contracts.
- Hyperledger Fabric is a permissioned blockchain that supports smart contracts.

3.14.4 Blockchain Platform Available over the Cloud

Several companies have offered Blockchain as a Service (BaaS) by allowing users to develop their own blockchain infrastructure using cloud. For instance, Microsoft Azure is one such platform.

3.15 Blockchain as a Service (BaaS)

Similar to other software platforms available through Software as a Service (SaaS) paradigm, vendors and service providers now offering blockchain platform as a service to the customers and to small and medium enterprises. This new paradigm is known as Blockchain as a Service (BaaS) in which the vendor and service provider are responsible for maintaining the blockchain infrastructure, while the customer can easily build their customized blockchain solution according to their requirements. Different service providers have initiated this step and provide blockchain service through their cloud platforms.

3.16   BitCoin Blockchain

Bitcoin is one of the famous first publicly available cryptocurrency platforms that use
the concept of blockchain. It uses Proof-of-Work (PoW) as consensus mechanism.
Bitcoin also has partial support of smart contracts using UTXO. In Bitcoin, everyone
can join the network, either as client or miner. Being a client, the participating Bitcoin
blockchain node can perform transactions (sending and/or receiving). The miner node
in Bitcoin blockchain network is responsible for mining process, i.e., to solve mining
puzzle through PoW. The Bitcoin blockchain operates over P2P network architecture.
Nodes first need to discover their neighbors through network discovery process. In
Bitcoin, there are also full nodes which keep the copy of the ledger. Bitcoin provides
a hard code list of Domain Name Server (DNS) IPs, which can be used together with
neighboring nodes to discover peers. In order to deal with DoS attack, Bitcoin links
the number of neighbor nodes to get connected.

In Bitcoin blockchain, virtual cryptocoins are used and these are referred to as Bitcoin (BTC). We can transfer BTC and record these transactions of Bitcoin blockchain.
The smallest amount of BTC that can be transferred is 0.00000001 BTC, which is
also known as "Satoshi".

$$1 \, Satoshi = \frac{1}{100000000} BTC, \tag{3.1}$$

These BTCs can be used to transfer assets "value" or it can be paid against the
mining process by the miners as a reward.

3.16.1   Creating Trust in Bitcoin Blockchain

Bitcoin creates trust without Centralized Authority using the following features:

- Cryptography.
- P2P network architecture.
- Hashing algorithms.
- Public/private key encryption.
- Distributed storage.
- Logically centralized.
- Consensus algorithm.
- Time stamping of Transactions.

3.16.2 Working of Bitcoin

In Bitcoin, when a blockchain participating node makes a transaction, it sends it to its neighboring nodes. The neighboring nodes first check this transaction, and if found it valid, send it to the network. In Bitcoin blockchain, transactions are executed sequentially. All the nodes including miners add this transaction to the unverified transaction pool. The miner node then gathers few transactions and starts mining the block. Once a miner successfully mined a block including set of transaction, it informs this to the whole network. The nodes (full nodes including miners) then once again check different fields of the block and check its validity. If found correct, then the nodes add this block to their copy of the ledger. In this manner, the transaction generated by the participating nodes becomes the part of the ledger.

3.16.2.1 Block

A block in a blockchain is composed of three main components: outer header, block header, and block body.

> The very first block created in a blockchain is known as "Genesis block".

3.16.2.2 Outer Header

The outer header consists of block size and block identification information. This also contains magic number. Block size mentions the size of the block, i.e., the maximum number in bytes.

> The magic number for bitcoin blockchain is 0xD9B4BEF9. This magic number is used for the identification of each block in the blockchain.

3.16.2.3 Block Header

The block header consists of information such as block version, time stamp, hashing target, nonce, parent block hash, and Merkle Tree root. All this information is useful in block validation.

3.16.2.4   Block Body

All the transactions are assembled in the Block body along with transaction counter.

3.16.2.5   Block Version Number

It specifies the version of blockchain protocol in use.

3.16.2.6   Parent Block Hash

This is the hash of the previous block to link different blocks in the blockchain.

3.16.2.7   Nonce

Nonce is used in validation process.

3.16.2.8   Time Stamp

Time stamp is used when a particular event happens.

3.16.2.9   Merkle Tree Root

Merkle Tree root contains the hash value. This hash value is generated using Merkle Tree procedure.

3.17   Ethereum Blockchain

Ethereum is another publicly available cryptocurrency. In Ethereum, Ethers are used as cryptocurrency. Ethereum brought innovation to the previously available Bitcoin blockchain in a sense that it incorporates business logic to blockchain network in the form of smart contract. Smart contracts can be written in Solidity programming language. Ethereum used Proof-of-Work (PoW) variant of consensus algorithm named as Ethash.

Full node in Ethereum needs 300 kb/s to run.

In Ethereum, every participating node needs to install/host Ethereum Virtual Machine (EVM) to execute smart contracts. Ethereum is also built on top of P2P network and requires virtual P2p (Vp2p) wire protocol.

It is worth noting that Ethereum can also support Proof of Authority (PoA) consensus protocol when used in private mode, i.e., as a private blockchain when installed in a private network. Thus, the transaction handling capacity of Ethereum is less than 50 transaction per second (tps) when used as a public blockchain and less than 1000 transaction per second (tps) when used as a private blockchain.

In Ethereum blockchain, there are three types of users: Contract Account, Miners, and Externally Owned Account.

1. Contract Account (CA), which are normal users and can make transactions among each other,
2. Miners, which are responsible for the mining process and
3. Externally Owned Account (EOA), which can perform transaction to another EOA. EOA can also call the function of CA and they can also make a new smart contract.

Ethereum is based on P2P, i.e., it uses Distributed Hash Table (DHT) as an underlying structure for its operations.

3.18 Hyperledger

Hyperledger is an open source project. This project is organized and managed by the Linux Foundation. It is the same initiative as open source operating system (Linux). In Hyperledger, different projects are on-going, each having a dedicated purpose. Table 3.4 shows projects managed under Hyperledger. Their code is available online and shown in Table 3.5.

3.19 Corda

R3 software company created the blockchain platform Corda. It is basically a permissioned blockchain and only permitted nodes can involve in the blockchain network. Corda is based on Hash-based Directed Acyclic Graph (Hash-DAG). Corda relies on X.509 certificate signing process for nodes. Corda can use RAFT consensus algorithm. Corda uses java virtual Machine (JVM) for its operation.

Table 3.4 Different projects managed under Hyperledger

Name of hyperledger project	Website link	Purpose
Hyperledger (Main Project)	https://www.hyperledger.org/	Umbrella project
Hyperledger Sawtooth	https://www.hyperledger.org/use/sawtooth	Distributed Ledger
Hyperledger Transact	https://www.hyperledger.org/use/transact	Library
Hyperledger Ursa	https://www.hyperledger.org/use/ursa	Library
Hyperledger Aries	https://www.hyperledger.org/use/aries	Library
Hyperledger Avalon	https://www.hyperledger.org/use/avalon	Tool
Hyperledger Besu	https://www.hyperledger.org/use/besu	Distributed Ledger
Hyperledger Burrow	https://www.hyperledger.org/use/hyperledger-burrow	Distributed Ledger
Hyperledger Cactus	https://www.hyperledger.org/use/cactus	Tool
Hyperledger Caliper	https://www.hyperledger.org/use/caliper	Tool
Hyperledger Cello	https://www.hyperledger.org/use/cello	Tool
Hyperledger Explorer	https://www.hyperledger.org/use/explorer	Tool
Hyperledger Fabric	https://www.hyperledger.org/use/fabric	Distributed Ledger
Hyperledger Grid	https://www.hyperledger.org/use/grid	Support Supply Chain—Distributed Ledger
Hyperledger Indy	https://www.hyperledger.org/use/hyperledger-indy	Distributed Ledger
Hyperledger Quilt	https://www.hyperledger.org/use/quilt	Library
Hyperledger Iroha	https://www.hyperledger.org/projects/iroha	Distributed Ledger

3.20   Tendermint

Tendermint is another blockchain platform based on the BFT consensus protocol. Tendermint basically relies on PBFT and PoS consensus protocols.

Table 3.5 Hyperledger projects and their corresponding codes

Name of Hyperledger project	Website link
Hyperledger Sawtooth	https://github.com/hyperledger/sawtooth-core
Hyperledger Transact	https://crates.io/crates/transact
Hyperledger Ursa	https://github.com/hyperledger/ursa
Hyperledger Aries	https://github.com/hyperledger/aries
Hyperledger Avalon	https://github.com/hyperledger/avalon
Hyperledger Besu	https://github.com/hyperledger/besu
Hyperledger Burrow	https://github.com/hyperledger/burrow
Hyperledger Cactus	https://github.com/hyperledger/cactus
Hyperledger Caliper	https://github.com/hyperledger/caliper
Hyperledger Cello	https://github.com/hyperledger/cello
Hyperledger Explorer	https://github.com/hyperledger/blockchain-explorer
Hyperledger Fabric	https://github.com/hyperledger/fabric
Hyperledger Grid	https://github.com/hyperledger/grid
Hyperledger Indy	https://github.com/hyperledger/indy-node
Hyperledger Quilt	https://github.com/hyperledger/quilt
Hyperledger Iroha	https://github.com/hyperledger/iroha

3.21 Chain Core

Another permissioned blockchain platform is Chain Core. It also uses UTXO model for transaction.

3.22 Quorum

Ethereum is a public blockchain and its permissioned version is called as Quorum blockchain. Quorum supports two consensus protocols, namely, RAFT and Quorum Chain.

> In majority of blockchain platform, consensus is reached at ledger level, while in Hyperledger, consensus is reached at transaction level.

3.23   Key Generation and Blockchain Digital Signature Procedure

In order to secure blockchain, every participating blockchain node generates a set of keys: public and private. The public key is shared among other nodes in the blockchain network in the form of hash code, generated from their public key. This hash code will serve as permanent address of the participating blockchain node and is used for identification purpose. This hash code (public key) is also termed as pseudo-identity of the participating blockchain node.

In blockchain, digital signatures are used to protect the blockchain. There are three steps involved in digital signature in blockchain.

1. Public/private key pair generation: When a user intends to generate a transaction, then it will first generate the public/private key pair to sign that transaction. Private key will be used to sign the transaction by the user and public key will be available publicly to decrypt the transaction.
2. Signing Phase: The next step is the signing. Before signing, the user needs to use hashing algorithm and then sign the transaction. Both public/private key generation and hashing are applied together in the data (transaction) to make it more secure.
3. Verification Phase: In this phase, the integrity of the blockchain data is verified. The user which received the encoded data (hashed and signed) will be decode by using the sender's public key. This data is then compared with the recomputed hash value. This will generate the original data. If both the hash values are same, it means the data received is original and unmodified.

3.24   Data Models in Blockchain

When a user wants to transfer an asset to another user, it can be done through transaction. Blockchain-based transaction has four basic data models: UTXO, Account, UTXO+, and key value. These different transaction data models have been adopted by different blockchain. Table 3.6 shows the blockchain data model along with its corresponding blockchain platform.

3.25   Implementation and Performance Evaluation Tools for DLTs

One of the important factors about implementation of blockchain system is related with its performance evaluation. Performance evaluation is important as it tells us which blockchain platform is better over the other in terms of various performance

Table 3.6 Data models used in Blockchain

Blockchain	Data model
Bitcoin	UTXO
Ethereum	Account
Corda	$UTXO^+$
Chain Core	$UTXO^+$
Hyperledger Fabric	Key value
Ripple	Account
IOTA	Account
Litecoin	UTXO

metrics such as throughput and latency. In addition, one can easily compare the performance of different versions of blockchain platforms. There are various ways to implement blockchain systems and evaluate its performance. These methods include mathematical modeling, simulation-based studies, and experimental deployments.

The performance of any blockchain system can be compared and evaluated using benchmarking tools such as Hyperledger Caliper, BlockBench, and DAGbench. These evaluating tools support several DLTs, the list of those are given below.

3.25.1 Hyperledger Caliper

Hyperledger Caliper is designed and developed by Hyperledger consortium. It supports the performance evaluation of various Hyperledger DLTs such as Fabric, Iroha, Sawtooth, and Besus. Hyperledger Caliper also supports Ethereum as well.

3.25.2 BlockBench

Blockbench is used for evaluating blockchain system. It supports Hyperledger, Quorum, Parity, and Ethereum. One can use different evaluation metrics such as throughput, latency, and fault tolerance of these blockchain systems. Blockbench is a performance evaluation platform in which private blockchain was evaluated and compared. The three private blockchain systems are Ethereum, Parity, and Hyperledger.

3.25.3 DAGBench

If one wants to evaluate the DAG-based distributed ledgers, then DAGbench is a good choice. It supports Nano, Byteball, and IOTA DAG-based DLTs for evaluation purpose.

3.25.4 How Consensus Algorithm Can Impact on the Performance of Blockchain?

Consensus algorithm can impact directly on the performance of blockchain, both in terms of transaction speed and finality of blocks. As we discussed, consensus algorithm is available in a variety of flavors, thus depending upon the underlying working of consensus algorithm, the performance of overall blockchain system changes. If the consensus algorithm is PoW, then the inter block time will be the time till the mining node wins the puzzle. It is approximately 10 min in Bitcoin blockchain. If the consensus algorithm is PoS based, then it totally depends on how blocks are mined. In some consensus algorithm, the role of leader is present, means a leader will be selected which will be responsible for mining process. Thus, the inter block time is directly related with the leader election/selection in the consensus algorithm. A more detailed discussion of consensus algorithm is present in Chap. 4.

3.26 Hashing in Blockchain

To secure blockchain and to achieve the inherent feature of immutability, blocks employ Hashing function. Hashing is applied at different levels in the blockchain. In Hashing, an input data is provided, and an output will be a fixed size output. For a particular input, the output will always remain the same. Even adding or modifying a single character in the input data will change the output value. This is that feature of hashing that is heavily used in the blockchain systems to make it secure and immutable. To summarize, hashing is used in blockchain to check the credibility, integrity, and authenticity of the data. Hashing is achieved by applying hash function to the input data. There are various hash functions available such as SHA224, SHA256, SHA384, SHA512, Blake2b, Blake2s, and MD5 Hashing algorithms. Just for illustration purpose we give an example of Hashing applied to Ethereum blockchain.

3.26.1 Hashing Applied to Ethereum Blockchain

Hashing is applied to calculate "parentHash", i.e., hash value of the parent block. Hashing is applied to calculate "ommersHash", i.e., hash value of the uncle block. When gathering transactions in a block, Merkle Tree is created. In Ethereum, Merkle Patricia Tree (MPT) is created in which hashing is applied. The result will be in the form of "TransactionRoot", i.e., hash root of the MPT by considering all transactions. Hashing is also applied to calculate "StateRoot" and "receiptsRoot". Root hash is calculated by using world state of MPT. Root hash is calculated by using all receipts.

Hashing is also applied when running "ethash". This consensus algorithm "ethash" is basically PoW consensus algorithm of Ethereum and calculates the hash value of the block in consideration. Moreover, miner nodes also need to solve the PoW puzzle and applying hashing extensively in the mining process. It is this execution of hashing function that requires heavy computation and results in a lot of energy consumption.

3.27 Data Storage in Blockchain

In a blockchain network, data can be managed and stored in two ways, either on the blockchain itself (On-chain) or only hashing information of important data is stored in the main blockchain and the other remaining data is stored offline (Off-chain). On-chain data management bring few advantages such as all the data can be accessible from the main blockchain, thus it increases transparency and blockchain records become auditable and thus tamper-resistant. On the contrary, Off-chain data management brings advantages that not huge amount of data is required to store on the blockchain, thus reducing storage requirements of blockchain nodes which is suitable for nodes which are not fully resourceful. Table 3.7 shows data management (On-Chain and Off-Chain) in a blockchain system.

Table 3.7 Data Management (On-Chain and Off-Chain)

	On-Chain	Off-Chain
Bitcoin	Yes	No
Ehtereum	Yes	No
Hyperledger	Yes	No
Tendermint	Yes	No

3.28 Data Structure in Blockchain

In a blockchain network, data can be stored and maintained using different data structures. These data structures can be organized in linear linked list. It can also be organized in the form of tree of blocks (as used in GHOST) or it can be Directed Acyclic Graph (DAG), as used in SPECTRE.

3.29 Privacy of Nodes in Blockchain

Privacy of nodes in blockchain network is managed through identity management system. If the identity of blockchain nodes is managed without a central server and blockchain nodes are not required to prove their identity by disclosing their personal information, then such type of identity management is called as self-sovereign identity system. This type of identity management is more private. If the identity of blockchain nodes is managed through the central server (also known as decentralized trusted identity management), i.e., at the very first time, blockchain node provides its personal information to prove its identity, then it is allowed further to generate the pair of further identities (keys) to blockchain nodes for identification purpose. This type of identity management requires the disclosure of personal information; thus, it is less private.

3.30 Smart Contracts

Smart contracts can be broadly classified into two major categories: deterministic smart contracts and non-deterministic smart contracts. If the execution of smart contract is not dependent upon external data or event, then such type of smart contract is referred to as deterministic smart contracts. If the execution of smart contract is dependent upon external data (also called as oracles), then such type of smart contract is referred to as non-deterministic smart contracts. Non-deterministic smart contract opened the door of new and innovative applications and business models. Below we mention few blockchain systems and discuss their smart contract feature.

3.30.1 Ethereum

Ethereum is public blockchain but it can be considered as consortium and private, as code is publicly available. Ethereum supports on-chain data management. Ethereum is based on PoW and it may switch to PoS.

3.30.2   Hyperledger

Hyperledger Fabric supports consortium and private blockchain. Non-deterministic smart contract feature is not available.

3.30.3   Tendermint

Tendermint supports public, consortium, and private blockchain.

3.30.4   Energy Web Chain (EW Chain)

This is a public consortium blockchain specifically designed to meet the requests of energy system and smart grid. EWF is based on Ethereum blockchain. Proof of Authority is the supported consensus algorithm and it supports both deterministic and non-deterministic smart contracts.

3.31   Scalability Issues in Blockchain Systems

One of the major issues in practically deployed blockchain system is scalability. Scalability can be considered in many aspects. For instance, how much scalable the blockchain system is when we increase the number of blockchain nodes? How blockchain system behaves when there is an increase in number of transactions? How blockchain system behaves when the size of the ledger increases in terms of storage? Basically, each blockchain keeps the copy of the complete ledger. For Ethereum, the complete ledger size is several hundreds GB.[1] This ledger contains the copy of all the transactions (blocks) from the current block to the very first block, i.e., genesis block. Note that with the passage of time, new blocks are adding, and the size of the ledger keeps growing. It is also important to consider that when applying blockchain to certain communication networks in which the nodes do not have high storage capacity, keeping the full copy of the ledger is problematic.

---

[1] Visit these websites for finding the current size of Ethereum blockchain:
https://etherscan.io/chartsync/chaindefault
https://blockchair.com/ethereum/charts/blockchain-size.

3.31.1 Blockchain Scalability Issues and Communication Networks

The major question with respect to blockchain scalability and communication networks is how the underlying communication network behaves and supports higher number of nodes and their generated traffic? This traffic is basically generated due to working principle of blockchain. In blockchain, when a blockchain node wants to include a transaction, it broadcast it to the whole network. This generates huge amount of communication overhead and it grows substantially when the number of nodes increases. Moreover, once the block is mined, the mined block is again propagated to the whole network. This further increases the communication overhead.

3.32 How to Increase the Transaction Capacity of Blockchain Systems?

The performance of blockchain system can be measured in terms of number of transactions handled per second by the blockchain. As mentioned earlier, Bitcoin can handle seven transactions per second (cf. Table 3.8 to see more details about transaction handling capacity of different blockchain system[2]). There are few factors that limit this transaction handling capacity. The first one is the number of transactions included in each block. The higher the number of transactions included in each block, the higher the throughput. Second, transaction speed is also dependent upon the underlying consensus algorithm, i.e., the mining process and the finality. If the mining process takes time, it will have a negative impact on the throughput. Thus, efficient and quick responding consensus algorithms are required. Third, the inter-arrival time of blocks will also have an impact on the transaction handling capacity of blockchain.

> Transaction handling capacity is one of the limitations of existing blockchain. For instance, roughly speaking, Bitcoin can handle less than ten transactions per second, Ethereum can handle less than hundred transactions per second, and Hyperledger can handle hundreds of transactions per second.

---

[2]Transaction rate per second can be found on this website:
https://www.blockchain.com/charts/transactions-per-second.

Table 3.8 Transaction handling capacity of different blockchain systems

Blockchain	Transaction handling capacity
Bitcoin	Can handle less than 10 Tx/s
Ethereum (in public setting)	Can handle less than 100 Tx/s
Ethereum (in private setting)	Can handle less than 1000 Tx/s
Hyperledger	Can handle hundreds of Tx/s

3.32.1   Off-Chain Transactions

Another possible way to increase the transaction capacity is to offload the transactions from the main blockchain to the side chain. This process is known as off-chain transactions. Researchers also explored the idea to store only the hashes in the main blockchain while other data need to store in off-chain solutions. In this manner, the overall storage requirement of keeping the ledger will not surpass the storage requirements of light weight nodes.

3.32.2   Sharding

Sharding is an important concept in blockchain which is proposed to handle transaction capacity, bottleneck of blockchain systems. In sharding, blockchain nodes are divided into multiple groups (shards) and each group (shard) is responsible to handle a certain amount of transactions in parallel. However, in sharding system, it is necessary to consider how all the blockchain nodes keep the copy of the complete ledger. Additionally, it is also important to consider how different blockchain nodes (group) communicate with each other.

3.33   Interoperability in Blockchain Systems

Consider a smart city environment in which multiple distributed ledgers are implemented and operated. Some of these ledgers are recording sensor information through IoT network, while other few ledgers are in place for management of land record. Other ledgers may be deployed for automobile industry and managing, operating, and recording the manufacturing, selling, buying, and registration of automobile and interacting with the revenue and registration departments of government. These ledgers can be of different architecture such as public, private, or consortium. However, there is a requirement that these ledgers "talk" with each other to establish an eco-system and facilitate the end users.

Another aspect is that these ledgers can be developed by different organizations and thus designed for specific purpose. Moreover, some of these ledgers are from 1st generation (not supporting smart contracts), while other ledgers are from 2nd generation (supporting smart contact). The most important point is that we cannot use a single ledger for all types of applications due to the availability of multiple ledgers in the market and these ledgers support various features. And still these ledgers are enhancing the performance. Thus, it is always complicated decision choice to select among these ledgers. The conclusive point is that interoperability is required among these ledgers to share information and talk with each other.

When we think about programming languages, interoperability solutions are present, and one may write code in one programming language and the program can interact with other languages. Similarly, in the context of traditional database systems, one may easily interact among several database systems, and migration of data as well as the data structure can easily be possible and interoperability feature is present in the context of programming languages and database systems.

To handle with this interoperability issue of different distributed ledgers, one may think about exchanges. Though exchanges are now worldwide present to share and exchange cryptocurrencies, but this is against the basic principle of public blockchain, i.e., requirement of trusted third party and reliance over it. Similarly, from the perspective of cryptocurrencies, these exchanges are present but what about sharing other assets and information among different blockchain systems and distributed ledgers?

3.33.1 Example to Understand Interoperability Issue

Ziad has an asset, for example, a car, recorded in a blockchain system. A smart contract is written by Zaid and it is mentioned that if Zaid dies, this car should be transferred to his friend Yasir only if Yasir does not have any agricultural land in his possession. Now, in this case, a smart contract has to be invoked and get information from the land record blockchain system and then this asset transfer can be completed. However, if Zaid's blockchain system was based on Hyperledger and Yasir's blockchain system is based on Ethereum, then what are the implications of this transfer? Will this be possible?

3.33.2 Using Smart Contract for Interoperability

Sharing information among different blockchain systems can be achieved through the use of smart contract. The smart contract is invoked on one blockchain system will interact with other blockchain systems to collect the required information and perform the required task. In this case, no trusted third party is required for interoperability among these different blockchain systems. However, it is essential that the blockchain system can easily understand the smart contracts.

3.33.3 Using Exchange for Interoperability

Another straightforward way to handle the interoperability among the blockchain system is to include an exchange, a controlling entity or a trusted third party, or even a separate blockchain system which records all such events and transactions. But again, this may violate the basic principle of public blockchain system.

3.33.4 Consensus Protocols and Interoperability Issue

Refer to our previous discussions about unique features of blockchain systems, we have identified that each blockchain system has its own characteristics that distinguish a particular blockchain from the other. Let us just take an example of consensus protocol (cf. Table 4.1), various ways were proposed to reach consensus in a blockchain system. Now the question is if a blockchain node that operates in PoW-based blockchain system can add a transaction to a PoS-based blockchain system? This arise the fundamental question that blockchain systems should be designed in such a manner or interoperability solutions are designed so that transaction among different blockchain systems having different consensus algorithms can easily be possible.

3.33.5 Interoperability Between Old and New Blockchain Systems

Generation of blockchain systems and interoperability among the newly designed blockchain systems is also required. For instance, Bitcoin blockchain does not support smart contract, while Ethereum blockchain does. When a blockchain user makes a transaction from Bitcoin to Ethereum then Ethereum can understand it but when Ethereum user with smart contract makes a transaction to Bitcoin blockchain then how Bitcoin blockchain system will react? In terms of underlying blockchain architecture, public blockchain systems enable any user to participate in the blockchain publicly, while in private or consortium blockchain, only limited blockchain users can participate after getting permission from the blockchain owners. In essence, public blockchain systems allow anyone to access the blockchain data, while this is not true in the case of private or consortium blockchain. Thus, question arises that when private or consortium blockchain data will be shared to the public blockchain system, how privacy can be achieved? Moreover, the transaction and data stored in private or consortium blockchain will be allowed to see in public blockchain in terms of blockchain user accessibility?

3.33.6 Transaction Speed and Interoperability

Transaction processing time or transactions per second is one of the important parameters through which we can analyze the performance of a blockchain system. This transaction processing time is dependent upon various factors among which consensus algorithm is a key factor. Some blockchain systems have higher transaction processing speed while others have lower transaction processing speed (cf. Table 3.8). Now imagine a scenario in which higher transaction processing system tries to add transaction to the lower transaction process blockchain system. In this case, what should happen? Faster blockchain should wait for the other blockchain system? If this is the case, then there will be a delay in updating both blockchain systems.

3.33.7 Semantic and Syntatic Interoperability

Blockchain systems store data in different formats when communicating among different blockchain systems, these formats need to be considered as well. Similarly, syntax of commands and semantics should also be considered when two or more blockchain systems communicate with each other. Thus, two types of interoperability, i.e., semantic and syntactic interoperability among the blockchain system, are required.

3.33.8 Transaction Fees and Interoperability

Besides transaction processing speed, each blockchain system charges its own transaction fees. Various fees are charged by the different blockchain systems. When a blockchain system sends a transaction to another blockchain system, the transaction speed needs to be considered in an interoperability solution, otherwise, it may affect the overall performance of blockchain system.

3.33.9 Tokens and Interoperability

In a blockchain system, an asset can be considered in the form of tokens. These assets can be a FIAT currency, a car, a land area, an agriculture land, a bond, or any

Table 3.9 Tokens in blockchain systems

Token	Type
Bitcoin	Bitcoin (native token)
Ethereum	Ether (native token)
FIAT money	Non-native token
Land piece	Non-native token
House	Non-native token
Car	Non-native token
Corda	Non-native token

other commodity. These tokens in the blockchain system can be classified as "native tokens" and "non-native tokens". Native tokens belong to a particular ledger such as Bitcoin blockchain has native token, i.e., Bitcoins, while Ethereum blockchain has native token called "Ethers". Non-native tokens can also be traded and exchanged in blockchain systems. In blockchain system, when exchanging these assets, interoperability among the assets, i.e., tokens also need to be considered. Table 3.9 shows different types of tokens in blockchain systems.

3.34   Summary

This chapter provided details on blockchain fundamentals and working principles. We first discussed blockchain network and types of blockchain networks. Categorization of blockchain was explained from different perspectives such as public, private, and consortium, and also from permissioned and permissionless perspective. We also discussed general issues with public blockchain system. Broadcasting and topology discovery were then discussed. We also highlighted different types of nodes available in blockchain system. Layers in blockchain system were also discussed in detail. We then discussed the general working sequence of blockchain. Different blockchain platforms were then discussed along with data models, hashing, data storage, and data structures. We also mentioned the implementation and performance evaluation platforms of blockchain. Finally, we mentioned the scalability and interoperability issues of blockchain systems.

3.35   Future Research Direction

One future research direction is how the number of nodes affects the consensus mechanism? Is there any dependency of number of nodes over the time to reach consensus? Any mathematical model is available to support?

3.36 Further Reading

For consensus protocols, readers are suggested to read this article [66]. A very detailed explanation of blockchain system and its history is presented in [18]. Scalability issues can be found in [98], while interoperability issues of blockchain systems can be found in [51]. Performance evaluation and experimental investigation of blockchain systems can be found in [29, 89].

Problems

3.1 You are required to explore how key generation and hashing can be carried out in Ubuntu Operating System. You should use the combination of basic Ubuntu OS commands to explore key generation and sha384 and md5 hashing algorithms.

3.2 Hyperledger is managed by which foundation?

3.3 Which companies offers to use Blockchain as a Service?

3.4 What are the benefits of using Blockchain as a Service?

3.5 How Microsoft's Azure differs from Hyperledger Fabric?

3.6 Explain the working of Hyperledger Fabric.

3.7 Which tools are used for the development of Hyperledger Fabric?

3.8 State few goals of Hyperledger.

3.9 What is the purpose of Hyperledger Composer?

3.10 State few Hyperledger services.

3.11 What are the differences between Bitcoin, and Ethereum?

3.12 What is the primary purpose of Bitcoin blockchain?

3.13 What is the difference between permissioned and permissionless blockchain?

3.14 Why one should use blockchain platform?

3.15 Define the role of Hyperledger working group?

3.16 Why modular design approach has been adopted in Hyperledger Fabric?

3.17 Describe the functionality of Fabric model.

3.18 State the steps involved in the development of blockchain network.

3.19 What is the purpose of Confidential Consortium Blockchain?

3.20 What is the vision of Microsoft's Azure?

3.21 How you will use Microsoft's Azure to develop blockchain network and use it as a platform?

3.22 Blockchain network is built on top of P2P network and it's working is based on propagation of transaction messages. How it is ensured that the messages are propagated and delivered successfully within the blockchain network?

3.23 What will happen if a communication link gets broken in a blockchain network?

3.24 What is the difference between tokens and cryptocurrency?

3.25 How to preserve the privacy of blockchain nodes?

3.26 How to achieve privacy in a blockchain network?

3.27 What is the structure of UTXO?

3.28 What are the contents of a transaction?

3.29 Explain Genesis block and how Genesis block of two blockchain systems differs?

3.30 Why and how Nonce is used in blockchain?

3.31 Write down the basic operations of blockchain.

3.32 What are the main participating nodes in blockchain?

3.33 How miners are incentivized in Bitcoin and Ethereum blockchain?

3.34 How new coins are maintained?

3.35 Differentiate between different types of blockchain systems.

3.36 Differentiate between public, private, and consortium blockchain.

3.37 Where Ethereum smart contract execute?

3.38 Explain the need of Ethereum Virtual Machine (EVM).

3.39 How transaction models are different in Bitcoin and Ethereum blockchain?

3.40 Smart contracts can be written in which programming language?

3.41 Which different accounts are present in Ethereum blockchain?

3.42 Draw the block structure for Bitcoin and Ethereum.

3.43 What is the role of miner node in public and private blockchain?

3.44 What is Ommer block?

3.45 What is the correct sequence of block creation in Bitcoin and Ethereum blockchain?

3.46 Why hashing is used in blockchain?

3.47 Write down two issues of symmetric key encryption.

3.48 Explain the working of public key encryption in your own words.

3.49 Define Elliptic Curve Cryptography (ECC).

3.50 In Ethereum, why hash functions is used?

3.51 What the requirements are of hash function?

3.52 Explain few differences between simple hash and Merkle tree hash.

3.53 How integrity of transactions is maintained in blockchain?

3.54 How addresses of accounts are generated in blockchain?

3.55 How complete transaction verification is carried out in blockchain?

3.56 How trust is established in blockchain?

3.57 What are the elements of trust in blockchain?

3.58 What is the difference between permissioned and permissionless blockchain?

3.59 Public blockchain is decentralized in nature, while consortium blockchain is partially centralized. Explain.

3.60 What are general issues in public blockchain?

3.61 How broadcasting is carried out in blockchain systems?

3.62 Differentiate between full node, lightweight node, and miner node.

3.63 What is the difference between leader node and validator node?

3.64 Blockchain system can be composed of how many layers? Explain the working of each layer.

3.65 What different data models are used in blockchain systems?

3.66 Who own blockchain? Support your answer with example.

3.67 1 Satoshi equals to how many BTC?

3.68 Assume a public blockchain network i.e., Bitcoin blockchain. Calculate the energy consumption of the entire network during the mining process. The current exchange rate of 1BTC = 40,798 Euros. The energy consumed during the mining process is measured in killoWatt-hours (kWH). Let's assume the block generation of Bitcoin network is 20 blocks per minute. The electricity unit price is 20.01 cents per kWh. Assume reasonable assumptions for transaction fees, and mining reward fees.

3.69 Which performance evaluation tools are available to evaluate blockchain systems?

3.70 What is the difference between on-chain and off-chain data management in blockchain?

3.71 What data structures are available in blockchain?

3.72 Name few scalability issues in blockchain system.

3.73 How transaction handling capacity of different blockchain systems can be increased?

3.74 Define sharding?

3.75 Why interoperability is important in blockchain systems?

3.76 Explain semantic and syntactic interoperability.

Chapter 4
Blockchain Consensus Algorithms

4.1 Consensus Algorithms

Blockchain can be viewed as a distributed database but what makes blockchain different from the traditional database? It is the quality of creating trust in a trustless environment and reach to consensus to make modifications or adding record to the blockchain, without a central entity. Another excellent feature of blockchain that distinguishes it from the traditional database is having the capability of tolerating nodes which somehow deviate from the global view of the blockchain itself. And this can go up to 51% of the nodes in the blockchain network (a.k.a., 51% attack). When such a huge number of nodes get compromised, then and only then blockchain data can be tampered easily. The core mechanism that reaches this decision (consensus) among the nodes is named as "consensus mechanism".

Consensus means to reach a common agreement. Consensus is required in different situations and it has several applications in our daily life. For instance, if the flight route requires an update, consensus must be reached among the involved and affected parties such as flight itself, and the control tower. Once this decision of consensus is reached, the consensus must be communicated to other parties and system must be updated. Consensus can be applied in different settings. It can be applied to a situation where all the participating entities are distributed and in order to reach a consensus, a protocol has to be followed. It can also be applied to a situation where we have a group of nodes which act as "central coordination committee", which is responsible to validate and enforce consensus. There can also be other settings as well where consensus can be adopted. In computer science, consensus has enormous applications. For instance, in Operating System (OS), and telecommunication systems, to handle any process or to deal with deadlock, consensus plays an important role.

When talking about consensus in a distributed environment, two types of nodes generally involved:

© Springer Nature Switzerland AG 2021
M. H. Rehmani, *Blockchain Systems and Communication Networks: From Concepts to Implementation*, Textbooks in Telecommunication Engineering,
https://doi.org/10.1007/978-3-030-71788-9_4

1. The legitimate nodes, which are the honest nodes, and
2. The malicious nodes, which act maliciously.

The system has to operate properly even if any failure occurs. There are two types of failures occur:

1. Crash failure, which occurs if legitimate node cause any failure, and
2. Byzantine failure, which occurs if malicious node causes any failure.

Blockchain systems are highly dependent upon consensus algorithms and the consensus algorithms have been studied for decades for a distributed environment. Thus, to understand well the behavior and working of consensus algorithms in a distributed environment, we discuss it later in this chapter, along with its features and properties.

One of the main responsibilities of consensus protocol is to maintain the data in the ledger (blockchain) in terms of ordering, originality, and tamper-resistant. Another main responsibility of consensus protocol is to reach consensus among the nodes in the blockchain network, i.e., to provide Byzantine agreement, even in the presence of misbehaving nodes. In this context, the consensus protocol is responsible for reaching a consensus of a particular state of the blockchain even when there are misbehaving nodes or different states of blockchain occur. This consensus protocol dictates which state of the blockchain is reliable and can be made available to all the nodes to save it as a local copy of the ledger.

In blockchain network, the trustless environment is present and blockchain nodes may behave maliciously or misbehave opposed to their original role. For instance, blockchain nodes may launch DoS attack or Sybil Attack or may perform double spending. Due to this misbehavior, different inconsistencies may occur in the blockchain itself, for instance, blockchain fork may occur (See Chapter 3 for details).

4.2 Functionality of Consensus Algorithm

The choice of selection of consensus algorithm is dependent upon the type of blockchain. If permissioned blockchain is considered, then the category of Byzantine Fault Tolerant (BFT) consensus protocol is suitable. If permissionless blockchain is considered, then a different category of consensus protocol is suitable. Figure 4.1 shows the classification of consensus protocols with respect to the type of blockchain and architecture, and Table 4.1 shows DLT and consensus algorithms used in these DLT.

In blockchain, we assume that few nodes exhibit Byzantine behavior and the majority of the blockchain nodes are honest. The number of Byzantine nodes in a blockchain consensus algorithm can support is one of the key requirements and assumptions considered when designing such systems.

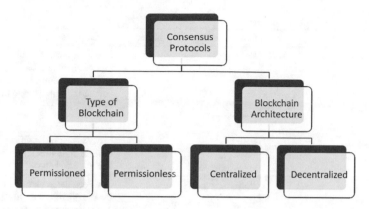

Fig. 4.1 Classification of consensus protocols with respect to the type of blockchain and architecture

Blockchain can also be viewed as data structure based on linear linked list of blocks tied together to form a blockchain. This linear linked list can store information such as transactions, time stamp, amount, and other information which users want to store in it. This linked list can grow infinitely and the integrity, and authenticity of this linked list will be verified and validated through "consensus" and "cryptographic" algorithms.

In order to update the state of the ledger (to add a block to the blockchain), consensus is required. There are four main responsibilities of the consensus algorithm:

1. Add a block to the blockchain.
2. Maintain the state of the blockchain.
3. Responsible to provide the same ledger to all the nodes, and,
4. Responsible to deal with data tampering in the blockchain from the malicious users.

4.3 Proof-of-Work (PoW) Consensus Algorithm

Proof-of-Work (PoW) is one of the famous consensus algorithms used in blockchain networks. It is the building block of Bitcoin blockchain and PoW is designed for public blockchain. In PoW, consensus finality is not guaranteed. It is worth noting that in PoW, the miner acts as both leader and validator node.

Consider a blockchain network of N participating blockchain nodes each having computational power $(\varphi_1, \varphi_2, \varphi_3, ..., \varphi_N)$ which can be used to solve the PoW puzzle. This computation power refers to the hash rate available to the participating blockchain nodes. The higher the hash rate, the higher the chances to solve the PoW

Table 4.1 DLT and consensus algorithms

DLT	Consensus algorithm Used	Description
Bitcoin	PoW	PoW operates in public blockchain and blockchain nodes are required to solve the cryptographic puzzle to win the mining process.
Ethereum	PoW	–
Hyperledger	PBFT	If 2/3 member confirm the block, then block become part of the blockchain.
Parity	PoS	In PoS, the selection of miners is dependent upon the amount of stakes each node carry instead of its computational power.
Hashgraph	Virtual voting-based consensus algorithm	In voting-based consensus algorithm, votes are required to reach to consensus.
Litecoin	PoW	–
Quorum	RAFT	–
Monax	Tendermint	–
ZCash	PoW	–
Hyperledger Sawtooth Lake	Proof of Elapsed Time (PoET)	Trusted and reliable hardware, such as Intel SGX is required to execute PoET consensus algorithm.
Tezos	PoS	–
Ripple	Ripple	–
IOTA	IOTA Tangle	–

cryptographic puzzle. The node that solves the PoW puzzle will be the winner and thus will be able to add block to the blockchain and deserve the reward. The winning probability P_w can be calculated as

$$P_{w_i} = \frac{\varphi_i}{\sum_{j=1}^{N} \varphi_j}, \tag{4.1}$$

where i is the participating blockchain node, N is the total number of nodes, and φ is the computation power of ith node.

In order to be consistent, we call this winner node as "leader node" because this will help us in explaining the working of other types of consensus algorithms.

PoW consensus is also susceptible to 51% attack. This 51% attack happens when 51% of the computational power is compromised and controlled by the attackers. This attack results in serious consequences for the blockchain network. For instance, attacker can control transactions or perform double spending.

PoW consensus has several issues. Double spending problem is another issue associated with PoW consensus algorithm. To address those issues, there are several consensus algorithms proposed to replace PoW consensus, which we are going to discuss next.

PoW is fault-tolerant to Byzantine Failure.

PoW is probabilistic in nature as two or more nodes may solve PoW puzzle simultaneously. This will result in the creation of "fork". In order to deal with forking, only that block is confirmed which is followed by six blocks in case of Bitcoin blockchain.

Even though PoW is designed for public blockchain but still PoW is not suitable for replacing the current banking system, as it requires to handle a large volume of transactions and PoW is probabilistic in nature.

PoW has a long latency issue, i.e., nodes take time to validate transactions.

4.3.1 Leader Node

A blockchain node which wins the mining process and receives the mining reward. In PoW, this winning will be dependent upon solving a cryptographic puzzle. In PoS, this winning is associated with stake.

4.3.2 Issues in PoW

PoW is computationally expensive. However, this computational expensive feature makes it attack resistant against Sybil Attacks, where malicious blockchain nodes create multiple pseudo identities to launch this attack.

4.3.3 How PoW Deals with Attacks?

In PoW consensus protocols, difficulty is one parameter which can be used to deal with attacks. When the difficulty level is high in the mining process, more computing resources are required to solve the mining puzzles; thus, it will be very hard for the attacker nodes to consume such high resources to continue their malicious activities. Another interesting fact in PoW based blockchain system is that the attacker needs to take over more than 51% of the blockchain nodes to make any changes in the blockchain which is practically hard considering the geographical distribution of blockchain nodes in a public blockchain.

4.3.4 Example of PoW Consensus Algorithm

Suppose a blockchain network with seven participating nodes (N_1, N_2, ..., N_7), having computing power (hash rate) values (φ_1, φ_2, φ_3, ..., φ_7). Calculate the probability of wining the puzzle by each participating node?

Solution: Let's assume the computation power (hash rates) are $\varphi_1 = 8$, $\varphi_2 = 4$, $\varphi_3 = 5$, $\varphi_4 = 10$, $\varphi_5 = 3$, $\varphi_6 = 1$, $\varphi_7 = 15$.

The winning probability P_w of each participating node can be calculated as

$$P_{w_i} = \frac{\varphi_i}{\sum_{j=1}^{N} \varphi_j}, \tag{4.2}$$

where i is the participating blockchain node, N is the total number of nodes, and φ is the computation power of ith node.

The wining probability of participating node N_1 can be calculated as

$$P_{w_1} = \frac{\varphi_1}{\sum_{j=1}^{7} \varphi_j}, \tag{4.3}$$

$$P_{w_1} = \frac{8}{8 + 4 + 5 + 10 + 3 + 1 + 15}, \tag{4.4}$$

$$P_{w_1} = 0.17, \tag{4.5}$$

Similarly, the wining probability of participating node N_7 can be calculated as

$$P_{w_7} = \frac{\varphi_7}{\sum\limits_{j=1}^{7} \varphi_j}, \tag{4.6}$$

$$P_{w_7} = \frac{15}{8 + 4 + 5 + 10 + 3 + 1 + 15}, \tag{4.7}$$

$$P_{w_7} = 0.32, \tag{4.8}$$

The winning probability of remaining participating nodes can be calculated in the similar manner.

4.4 Proof of Stake (PoS) Consensus Algorithm

In Proof-of-Stake (PoS) category of consensus protocol, the selection of miners is dependent upon the amount of stakes each node carries instead of its computational power. The higher the stake a node has, the higher the chances to select that node as a winning miner node. One of the advantages of PoS category of consensus protocol over PoW is less consumption of energy. Secondly, in PoS consensus algorithm, nodes do not need to be as much powerful in terms of computation power (hardware) as compared to nodes which run PoW. In terms of transaction confirmation speed and block generation, the PoS category of protocol has a faster speed as compared to PoW. This is primarily due to single block generation in each round of PoS.

> PoS can be considered as performing "virtual mining", i.e., no energy consumption is required as needed in PoW.

In PoS consensus, the winner is the one which has a higher stake. These stakes can be in any form. It can be the amount of cryptocurrency a node owns or it can be the amount of digital token a node carry or it can be the amount of energy a node has or it can be the amount of computational power a node has. Follow-the-Satoshi (FTS) algorithm has been used in many PoS consensus implementations. In FTS, a seed is provided as an input and then hashing is applied to it. As an output, the FTS algorithm gives a token index. By using this token index, the FTS algorithm nominates the miner which has this same token index by using the transaction history.

4.4.1 Issues in PoS

Network links and connectivity play an important role in the performance of PoS consensus protocols. If the messages are communicated to the nodes in the network without loss and delays, the block verification can be quickly done, and nodes update the more recent copy of the ledger on their systems. However, if there is any delay or link connectivity issues, nodes may not have the recent copy of the ledger and thus, results in the synchronization problem.

Another issue related to network link and connectivity is associated with PoS consensus algorithm is the message communication among the committee members of some PoS variants. This includes the voting decision as well as committee head notification. Without timely sharing of messages among the committee members, the performance of PoS consensus algorithm can be seriously jeopardized.

As highlighted previously, incentivization is important to sustain the blockchain. In this context, rewarding the winning node is important. In PoS, reward should be awarded in such a manner that all the nodes get equal opportunity to participate in the mining process, otherwise, a situation will arise when only the mining node which has the higher stake will always get rewarded and always wins the mining process. This situation will be monopolized by a single node and this needs to be considerd when designing future consensus algorithms.

4.5 Mining Pools

Considering the context of cryptocurrency such as Bitcoin and Ethereum, participating nodes get reward from the mining process. The participating nodes which have higher computational power will have higher chances to win the mining process, while nodes which will have less computational power will have less chances to win the puzzle. A natural tendency will be in this case is to join the mining pool by the participating nodes to share their computational power and jointly solve the puzzle. Of course, mining pools will have more resources (computational power) than the individual nodes and thus have higher chances to win the puzzle than the individual nodes. Mining pools also consume lot of energy.

This natural tendency of nodes joining the mining pools will lead to a situation where mining pools will be dominating the mining process. If we see the current state of mining process (as of 21st November 2020 for a period of four days), few mining pools are dominating and their percentage of mining is quite high (cf. Figure 4.2). Table 4.2 shows the name of mining pools along with the number of blocks mined by them (as of 21st November 2020).[1]

---

[1] Data taken from the website: https://www.blockchain.com/pools.

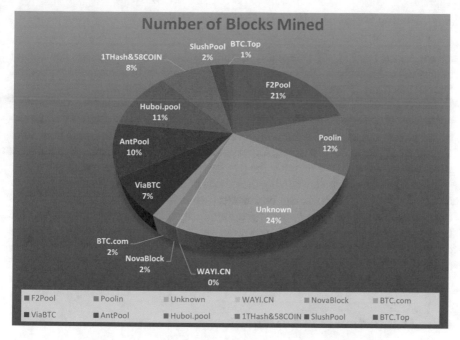

Fig. 4.2 Pie Chart showing the percentage of blocks mined by different mining pools

Table 4.2 Mining pools and their corresponding blocks mined

Mining pools	No. of blocks mined
F2Pool	134
Poolin	72
Unknown	148
WAYI.CN	1
NovaBlock	9
BTC.com	11
ViaBTC	46
AntPool	62
Huboi.pool	71
1THash&58COIN	53
SlushPool	14
BTC.Top	9

4.6   Issues Related with Mining Pools

As can be seen in Fig. 4.2 (Pie chart) that mining pools are dominating and adding blocks to the blockchain. This is against the key feature of blockchain networks (cf. Sect. 1.2 in Chap. 1) and it is more specific to Bitcoin blockchain.

Due to the distributed and decentralized nature of Bitcoin blockchain, performing consensus is the responsibility of miners. These miners have different computational power and some of them collaborate in mining pools to increase their joint computation power. In few cases, it may possible that different miners at different geographical location try to solve the puzzle in order to gain incentive. This may result in two or more miners solving the puzzle at the same time. As a result, two or more versions of blockchain can be created (see more details in Chapter 3). Moreover, since the blockchain network is geographically distributed and operate over the Internet infrastructure; therefore, some node may have healthy and fast links than the others. Additionally, propagating the blocks to reach the whole network may further cause the delay.

4.7   Transaction (Tx) Throughput

The basic component of blockchain is Transaction (Tx). These transactions are assembled to form a block. Different blocks are linked together to form the blockchain. To maintain the integrity and verifiability of blockchain, cryptographic algorithms will be applied at different stages (more details are discussed in Chapter 3).

Transaction throughput is defined as the number of transactions per second a blockchain network can process. It is measured in Tx/s, where Tx is the number of transactions and 's' means second. It can be calculated as

$$Transaction\ Throughput = \frac{Block\ Size\ (Bytes)}{Transaction\ Size\ (Bytes) \times Block\ Time\ (seconds)},$$
$$(4.9)$$

where "Block Size" is the size of the block and varies from blockchain to blockchain. For instance, Bitcoin has 1 MB Block size. "Transaction size" is the size of a transaction and it can vary from transaction to transaction and from blockchain to blockchain. For Bitcoin, it is 250 Bytes.

And "Block Time" is the time required to add a block to the blockchain. It is considered as average, as block time varies from block to block. In simple words, Transaction Throughput tells us the time required to add a transaction to the blockchain network.

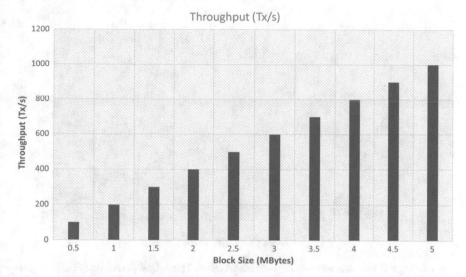

Fig. 4.3 Impact of block size (MB) over Throughput (Tx/seconds). Here, we kept transaction size is 500 Bytes and block confirmation time is 10 s

4.8 Block Confirmation Time

It is defined as the time to confirm the block and it tells us how quickly a block can be confirmed in a blockchain. Depending upon the underlying blockchain, it can further be dependent upon the number of blocks. A new block must wait before it is confirmed. It can be measured in seconds. It can also be referred to as finality.

4.9 Impact of Tx Throughput and Block Size

Figure 4.3 shows the impact of block size (MB) over Throughput (Tx/seconds) where we kept transaction size is 500 Bytes and block confirmation time is 10 s. It can be observed that the Throughput (Tx/seconds) increases linearly with the increase in block size (MB). The primary reason for this increase in Throughput is due to higher number of transactions supported by large block size.

4.10 Impact of Block Confirmation Time and Throughput

Figure 4.4 shows the impact of block confirmation time (seconds) over Throughput (Tx/second). Here, we kept block size is 1 MB Bytes and Tx size is 500 Bytes. It can be observed that the Throughput (Tx/second) decreases with the increase in block confirmation time (seconds). The primary reason for this decrease is due to decrease in addition of block to the blockchain.

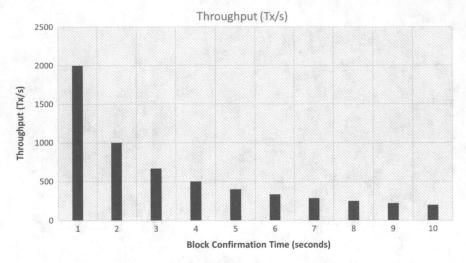

Fig. 4.4 Impact of block confirmation time (seconds) over Throughput (Tx/second). Here, we kept block size is 1 MB Bytes and Tx size is 500 Bytes

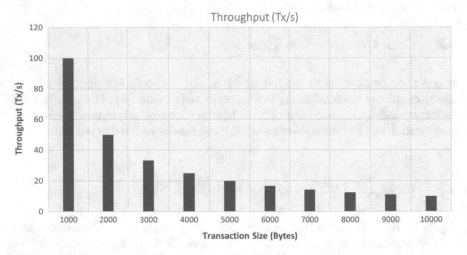

Fig. 4.5 Impact of transaction size (Bytes) over Throughput (Tx/second). Here, we kept block confirmation time is 10 s and block size is 1 MB

4.11 Impact of Transaction Size and Throughput

Figure 4.5 shows the impact of transaction size (bytes) over Throughput (Tx/second). Here, we kept block confirmation time is 10 s and block size is 1 MB. It can be observed that the Throughput (Tx/second) decreases with the increase in transaction size (bytes). When we have big size of the transactions then it will result in less number of transactions available for validation in this block and this results in decrease in Throughput.

4.12 Example of Tx Throughput and Block Confirmation Time

Suppose a blockchain network has a block size of 2 MBytes, transaction size of 500 Bytes, Block time is 200 s and it requires to wait for 10 blocks to confirm any new block. Calculate average block confirmation time and transaction per second (throughput) of this blockchain network?

$$Transaction\ Throughput = \frac{2\ (MBytes)}{500\ (Bytes) \times 200\ (s)}, \tag{4.10}$$

$$Transaction\ Throughput = \frac{2 \times 10^6}{500 \times 200}, \tag{4.11}$$

$$Transaction\ Throughput = 200\frac{Tx}{seconds}, \tag{4.12}$$

$$Average\ Confirmation\ Time = 10 \times 200\ s = 2000\ s, \tag{4.13}$$

The Transaction Throughput is 200 Tx/s and Average Confirmation Time is 2000 s.

4.13 Different Consensus Algorithms

In this section, we outline different consensus algorithms.

4.13.1 Proof-of-X

Proof-of-X category algorithms can scale up well and work well in public blockchain networks. However, the BFT category of consensus protocols does not scale well and are thus more suitable for permissioned blockchain environments.

4.13.2 Hyrid Consensus Protocol

A new type of consensus protocol is also possible, i.e., to combine both the features of PoX and BFT protocols, thus, creating a "hybrid" of consensus protocols. In this hybrid category of consensus protocols, "block generation" is the responsibility of one type of consensus protocol, and "block validation" is carried out by another type of consensus protocol.

4.13.3   PoW-PoS Protocols

In this type of consensus protocol, the consensus protocol keeps the feature of both PoS and PoW protocols. One such example is the Peercoin consensus protocol. Proof-of-Activity (PoA) is another example of consensus protocol. In PoA, empty blocks are created using PoW, and block verification and transactions are added using PoS. Snow White is another protocol which uses both PoS and PoW, in which the first PoS is used for the selection of candidates and then PoW is used for the creation of blocks.

4.13.4   Committee-Based Consensus Algorithms

Another type of consensus protocol is based on committee members for block verification, block addition, and addition of transactions. The selection of these committee members can be based on certain criteria. Peersensus is a consensus protocol which uses this concept of committee. Hybrid consensus protocol is another such protocol which uses the same concept. The energy consumption of committee-based consensus protocol is much higher than PoS algorithm, however far less than PoW consensus. The block generation speed and transactions confirmation speed of committee-based consensus protocol are faster than PoW and slower than PoS.

4.13.5   Consensus Protocols for Distributed Data Storage

In order to select a successful miner, nodes must show proofs of storage. The higher the storage possess by the node, the higher the probability to win the mining puzzle. Examples of such algorithms are Permacoin, Koppercoin, and Filecoin.

4.13.6   Proof-of-Human-Work

Proof-of-Human-Work employs Completely Automated Public Turing Test to tell Computers and Humans Apart (CAPTCHA).

4.13.7   Primecoin

Primecoin is a consensus algorithm which require the participating node to search for three types of prime number chains.

4.13.8   Proof-of-Exercise

Proof-of-Exercise solves matrix product problem to win the mining process.

4.13.9   Proof-of-Useful-Work

Proof-of-Useful-Work solve useful functions to win the mining process.

4.13.10   Ouroboros Conesus Protocol

Ouroborous consensus protocol is a type of PoS consensus which was developed in 2017. It uses Follow-the-Satoshi (FTS) algorithm. When the miner nodes are interested to mine a block, Ouroborous consensus algorithm establishes a committee. This committee is responsible to select a wining miner node among the miners to add a block to the blockchain.

It is now used by the cryptocurrency Cardano and sp8de. Ouroborous divides the physical time into epochs and each epoch is further divided into N slots. During this phase, the committee members select the wining node. In Ouroborous, a winning miner node is selected for each epoch.

The selection of committee is dynamic in Ouroborous. In Ouroborous, the wining miner node only creates empty blocks, and transactions are only added to these blocks by the input endorsers.

4.13.10.1   Attack Resilience of Ouroborous

Ouroborous is attack resilient to different attacks such as grinding attack, long-range attack, and nothing-at-stake attack.

4.13.11   Chain of Activity

Chain of Activity (CoA) consensus algorithm is also a type of PoS consensus. It follows FTS algorithm to select the winning miner node. However, the main difference between CoA and Ouroborous is that in CoA, the seed input to the FTS algorithm is generated using hash of the blocks created in each epoch. Basically, in CoA, the chain of blocks created is divided into groups and each group has a length of blocks. Then all the blocks created in each group are used to generate the seed. First, a seed of each block is generated using the block's hash. Then all the seeds are combined to input the FTS algorithm. CoA consensus algorithm is used in Tezos cryptocurrency.

4.13.12   Casper

Casper is another PoS based consensus protocol. It basically employs PoW, however, a voting mechanism is introduced. This voting is designed in such a manner that the voting power is proportional to the stake hold by the mining nodes. Casper is used in Ethereum blockchain.

4.13.13   Algorand

Algorand is another PoS consensus protocol. It is used in cryptocurrencies such as Algorand and Arcblock. Instead of using FTS algorithm, the Algorand algorithm uses cryptographic sortation mechanism to select the committee members. This committee is responsible to nominate the winning node in a round-robin fashion. However, the winning node cannot directly add the new block to the blockchain, instead, the winning node passes this new block to the committee, and then the committee is responsible to add this block (finalize) to the blockchain. In Algorand, depending upon the amount of stake, each node holds a range of hash values assigned. Algorand's committee generates the hash value. If this hash value matches any of the assigned hash value, the corresponding node will be declared as the winning minner.

Algorand is also resistant to several attacks such as grinding attack, double spending, bribe attacks, nothing-at-stake, and long-range attacks.

4.13.14 Tendermint

Tendermint is another PoS based consensus algorithm. The committee is responsible to nominate the block to be added to the blockchain. Tendermint is used in
BigchainDB and Ethermint.

4.14 Consensus Protocol for Permissioned Blockchain

In a permissioned blockchain, only specific miner nodes (a group) can perform
the consensus mechanism with tighter control in the mining process. In this context,
specific consensus protocols are applicable, for instance, the Byzantine Fault Tolerant
(BFT) consensus protocol. In practical blockchain implementations, Ripple, and
PBFT are examples of such types of protocols.

4.15 Consensus Protocol for Permissionless Blockchain

In a permissionless blockchain (public blockchain), any node can participate in the
mining process. Moreover, reaching consensus in this completely decentralized public blockchain is a bit challenging, as the environment is completely trustless and
there are no centralized nodes to manage the consensus process. Additionally, any
node can join or leave the public blockchain, thus, scalability is one of the major
issues in such blockchain networks.

4.16 Why BFT Protocols Cannot Be Used in Public
Blockchain?

As mentioned earlier, public blockchain has a trustless environment and there is no
control on the joining/leaving of nodes. Nodes typically use pseudonymity (i.e., their
actual identities are not disclosed), thus identity authentication in a public blockchain
is a difficult task. In this context, BFT consensus protocol is not suitable as they
require voting, which is credible when the identity of voting nodes is authenticated.
Thus, in the case of public blockchain, PoW consensus protocol has been widely
adopted because it requires PoW node to solve cryptographic puzzle and incentivize
nodes to take part in the mining process.

4.17 Summary

This chapter discussed in detail the core component of any blockchain system, i.e., consensus algorithms. First, the definition of consensus algorithm is presented and then we discussed the functionality of consensus algorithm. We further discussed state-of-the-art consensus algorithm used in various blockchain systems. We also discussed how consensus algorithm deals with attacks.

4.18 Further Reading

This section outlines further reading material for consensus algorithms. There are several survey articles published recently on consensus algorithms [66, 92, 97]. To read more about data processing view of blockchain systems, the article [26] is a wonderful resource.

Problems

4.1 Suppose a blockchain network has a block size of 4 MBytes, transaction size of 100 Bytes, Block time is 400 seconds and it requires to wait for 20 blocks to confirm any new block. Calculate average block confirmation time and transaction per second (throughput) of this blockchain network?

4.2 Suppose a blockchain network with seven participating nodes ($N_1, N_2, \ldots, N_7$), having computing power (hash rate) values ($\varphi_1, \varphi_2, \varphi_3, \ldots, \varphi_7$). The computation power (hash rate) are: $\varphi_1 = 18$, $\varphi_2 = 14$, $\varphi_3 = 15$, $\varphi_4 = 100$, $\varphi_5 = 30$, $\varphi_6 = 16$, $\varphi_7 = 150$. Calculate the probability of wining the puzzle by each participating node?

4.3 What are the connectivity requirements of consensus protocols in blockchain network?

4.4 What is fault tolerance and why it is important in the context of consensus protocols?

4.5 How to measure the strength of a consensus protocols?

4.6 What is double spending problem and how it can be addressed?

4.7 What exceptions can occur in blockchain network?

4.8 How we can differentiate between hard fork and soft fork?

4.9 What is the difference between crash failure and Byzantine failure?

4.10 What is the main functionality of consensus algorithm?

4.11 Explain the working of proof-of-work consensus algorithm?

4.12 Explain the need of proof-of-stake consensus algorithm.

4.13 What are the issues in PoW consensus algorithm?

4.14 Explain the role of mining pools and its impact on mining process.

4.15 Explain how PoW deals with attack?

4.16 What are the issues in PoS?

Part II
Hands-on Exercises and Blockchain Implementation

This part of the book consists of one chapter (chapter five) in which two mini projects are presented. Moreover, this chapter also contains five lab implementations along with desired program output and sample code.

Chapter 5
Hands-On Exercise and Implementation

In this chapter, we discuss few mini projects, hands-on exercises, and implementation code related to blockchain. These mini projects, hands-on exercises, and lab implementation codes provide the basis to further build code for blockchain systems for communication networks.

5.1   Mini Project 1: Critical Analysis of Distributed Ledger Technology

This mini project is worth 40 marks. It requires you to work on any of the following topics about distributed ledger technology, critically analyze it, and also discuss future research directions based upon state-of-the-art work.

Consider any of the following use case scenarios:

- Supply Chain
- Smart Factory
- Energy Trading in Smart Grid
- Pharmaceutical Industry
- Auto-mobile Industry
- Real Estate Transactions

Note: It is important that you should not copy any existing use case scenario and mention it in your assignment. For instance, if you go to Hyperleger website or IBM blockchain website or see any published research paper, they have mentioned the complete architectures for different blockchain use case scenarios. Simply copy/paste those architectures in your project will not be a good idea. However, you can get some insights from those scenarios and propose your own architecture for any of the above use case scenarios.

© Springer Nature Switzerland AG 2021
M. H. Rehmani, *Blockchain Systems and Communication Networks: From Concepts to Implementation*, Textbooks in Telecommunication Engineering,
https://doi.org/10.1007/978-3-030-71788-9_5

5.1.1 Questions

1. Give short background and draw the architecture of the proposed blockchain.

   [2 Marks]

2. Describe the functionality of each component in your proposed architecture.

   [2 Marks]

3. Which blockchain (Public, Private, Consortium) you will select and why?

   [1 Marks]

4. Your blockchain architecture will be Permissioned or Permissionless? Discuss in detail.

   [1 Marks]

5. Which Hashing algorithm you will use and why?

   [1 Marks]

6. How you will ensure trust in your proposed blockchain Architecture?

   [2 Marks]

7. Which Consensus algorithm you will use and why? Explain its working as well.

   [3 Marks]

8. What changes are required in your proposed architecture if we use Proof of Work (PoW), Proof of Stake (PoS), Practical Byzantine Fault Tolerance (PBFT), and Proof of Elapsed Time (PoET) consensus algorithms?

   [3 Marks]

9. In your proposed blockchain architecture, how Blocks will be validated, verified, and added?

   [2 Marks]

10. How you will deal with double spending problem in your proposed blockchain architecture?

   [2 Marks]

11. How smart contracts will be integrated and executed in your proposed blockchain architecture? Give two example of using smart contracts in your selected scenario.

   [2 Marks]

12. Propose any DApp for your considered scenario.

   [2 Marks]

13. Which evaluation metrics (describe any two) can be used to evaluate the performance of your proposed blockchain architecture?

[2 Marks]

14. What will be the outcome (two benefits) if the proposed blockchain architecture is adopted?

[2 Marks]

15. Describe three future research directions in relation to your proposed blockchain architecture. These future research directions should be discussed in detail. Technical depth of discussion is required.

[3 Marks]

5.2 Mini Project 2: Implementation of Distributed Ledger Technology and It's Security Analysis

This mini project is worth 60 marks. A topic about distributed ledger technology is given to you; you need to implement it using open source programming language such as Python. You are also required to critically assess the performance of variants of PoS consensus algorithms and also critically analyze the security challenges and perform threat analysis.

Consider any of the following use case scenarios or you may consider your own scenario:

- Supply Chain
- Smart Factory
- Energy Trading in Smart Grid
- Pharmaceutical Industry
- Auto-mobile Industry
- Real Estate Transactions

Note: There are total three questions in this mini project 2 which are required to solve. For the threat assessment modeling, the following article [42] is a good resource.

Cedric Hebert and Francesco Di Cerbo, "Secure blockchain in the enterprise: A methodology", Pervasive and Mobile Computing, Volume 59, October 2019, 101038.

https://www.sciencedirect.com/science/article/pii/S1574119218307193

It is important that you should not copy any existing use case scenario and mention it in your assignment. For instance, if you go to Hyperleger website or IBM blockchain website or see any published research paper, they have mentioned the complete architectures for different blockchain use case scenarios.

Simply copy/paste those architectures in your assignment will not be acceptable. However, you can get some insights from those scenarios and propose your own architecture for any of the above use case scenarios. Moreover, do not copy/paste any available code on github. You need to write the code by yourself.

5.2.1 Questions

Part A
[Total Marks = 35]

1. Give short background and draw the architecture of the proposed blockchain.

[2 Marks]

2. Use the article [42] and evaluate the security of your proposed blockchain architecture by using Microsoft's STRIDE methodology. You need to follow the same methodology as mentioned in the aforementioned article, i.e., try to follow the four main steps and also answer the 67 blockchain-specific threats.

   a. Step 1: Discuss whether blockchain is an applicable solution for your proposed blockchain. Here answer the seven questions mentioned in the paper.

[7 Marks]

   b. Step 2: Critically analyze your selected blockchain implementation.

[11 Marks]

   c. Step 3: You need to mention your analysis for sixty blockchain specific threats for your proposed blockchain (cf. [42] for more details). Please write the name of the threat while answering it.

[12 Marks]

   d. Step 4: For your proposed blockchain solution, try to perform threat assessment modeling.

[3 Marks]

Part B
[Total Marks = 25]

In Part B, 15 Marks is for Implementation of Code in Python and 10 Marks is for Comparison through graphs showing Mining time and throughput.

3. Implement and compare the performance of any of the following three Proof-of-Stake Consensus algorithms. You need to demonstrate which of the PoS algorithm is taking more time in Mining, and also show throughput (in terms of transaction per second (Tps)).

a. Ouroboros PoS Algorithm
b. Chain of Activity
c. Algorand
d. Tendermint
e. Casper

5.3 Lab Implementation 1

5.3.1 Aim

Hashing is one of the building blocks of any blockchain implementation. The aim of this lab is to understand how Hashing works. More specifically, the aim is to design an application in Python which performs Hashing and check generated hash values. Few famous Hashing algorithms that are supported by Python programming language (Hashlib and zlib) are:

- MD5 Hashing Algorithm – md5()
- Secure Hash Algorithm 1 (SHA-1) – sha1()
- Secure Hash Algorithm 2 (SHA-2) Family: This family of SHA-2 algorithm consists of the following Hashing variants.

    - SHA-224 – sha224()
    - SHA-256 – sha256()
    - SHA-384 – sha384()
    - SHA-512 – sha512()

- Blake2 Hashing Algorithm: This family of Blake2 algorithm consists of the following two Hashing variants.

    - Blake2b – blake2b()
    - Blake2s – blake2s()

- Secure Hash Algorithm 3 (SHA-3) Algorithm Family: This family of SHA-3 algorithm is subset of Keccak algorithm. It has the following Hashing variants.

- SHA3-224 – sha3_224()
- SHA3-384 – sha3_256()
- SHA3-384 – sha3_384()
- SHA3-512 – sha3_512()
- Shake-128 – shake_128()
- Shake-256 – shake_256()

- Adler32 is basically a checksum – adler32(). This function is available in the zlib module.
- CRC32 is also a checksum – crc32(). This function is available in the zlib module.

5.3.2 Steps to Follow

You are required to perform these steps:

- First, import the Python libraries. This include the hashlib. If you want to use Adler32 and CRC32 then you also need to import zlib as well.

  - Hashlib: This library will be used to create hash for every block. Hashlib has several built-in functions which you can use. For instance, sha1(), sha224(), sha256(), sha384(), sha512(), and blake2s(). Moreover, the reason for using sha256() is that it has been used in Ethereum and Bitcoin blockchains.

- Create a loop to iterate number of times user want to check the number of Hashes.
- Ask user to provide data to generate its Hash Value.
- You should also demonstrate that if the data input by user is same, the generated hash value will be same.

5.3.3 Desired Program Output

Output of the Program
The output of the program is shown in Fig. 5.1.

5.3.4 Sample Code

```
1  # Lab 1 program
2
3  import datetime
4  import hashlib
```

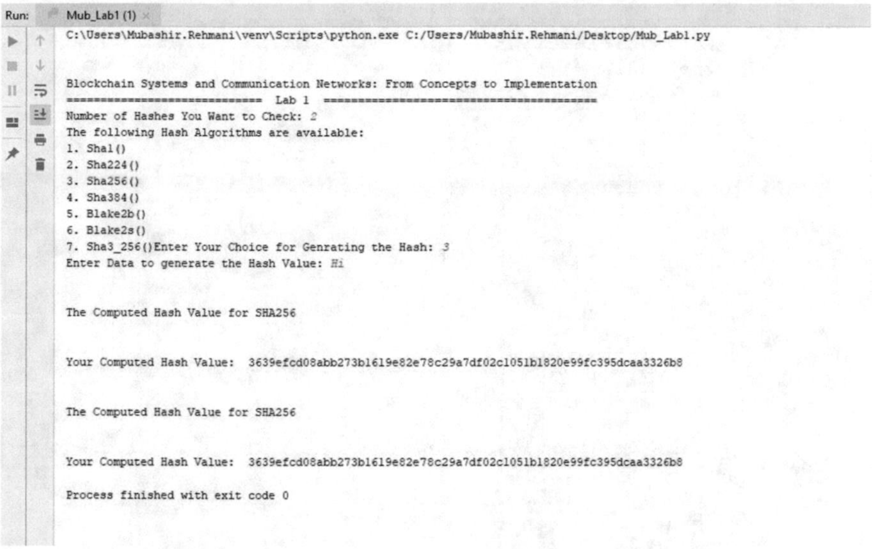

Fig. 5.1   Out of the program: lab1

```
5
6  print("\n")
7  print("Blockchain Systems and Communication Networks: From
      Concepts to Implementation")
8  print("============================   Lab 1
      ======================================")
9
10 loopnmbr = int(input('Number of Hashes You Want to Check: '))
11 typeofhash = int(input('The following Hash Algorithms are
      available:\n1. Sha1() '
12                        '\n2. Sha224() \n3. Sha256() \n4. Sha384()
      \n5. Blake2b() '
13                        '\n6. Blake2s() \n7. Sha3_256()'
14                        'Enter Your Choice for Genrating the Hash:
      '))
15 data = str(input('Enter Data to generate the Hash Value: '))
16
17
18 class Lab1_Hashing():
19
20
21    def hash(self):
22        string = data
23        string = string.encode('utf-8')
24        if typeofhash == 1:
25            h = hashlib.sha1(string)
26            print("\n")
27            print("The Computed Hash Value for SHA1")
28            return h.hexdigest()
29        if typeofhash == 2:
30            h = hashlib.sha224(string)
31            print("\n")
```

```
32              print("The Computed Hash Value for SHA224")
33              return h.hexdigest()
34          if typeofhash == 3:
35              h = hashlib.sha256(string)
36              print("\n")
37              print("The Computed Hash Value for SHA256")
38              return h.hexdigest()
39          if typeofhash == 4:
40              h = hashlib.sha384(string)
41              print("\n")
42              print("The Computed Hash Value for SHA384")
43              return h.hexdigest()
44          if typeofhash == 5:
45              h = hashlib.blake2b(string)
46              print("\n")
47              print("The Computed Hash Value for Blake2b")
48              return h.hexdigest()
49          if typeofhash == 6:
50              h = hashlib.blake2s(string)
51              print("\n")
52              print("The Computed Hash Value for Blake2s")
53              return h.hexdigest()
54          if typeofhash == 7:
55              h = hashlib.sha3_256(string)
56              print("\n")
57              print("The Computed Hash Value for SHA3_256")
58              return h.hexdigest()
59
60
61  def main():
62
63      for n in range(loopnmbr):
64          hashes = Lab1_Hashing()
65          output1 = hashes.hash()
66          print("\n")
67          print("Your Computed Hash Value: ", output1)
68
69  main()
```

5.4   Lab Implementation 2

5.4.1   Aim

The aim of this lab is to include time stamp in the data provided by the user. In this manner, each time if the same data is provided by the user, it will generate a different Hash value. This functionality is important as time stamps are included in every block in the blockchain.

5.4.2   Steps to Follow

You are required to perform these steps:

- First, import the Python libraries. This includes the datetime library and hashlib. If you want to use Adler32 and CRC32 then you also need to import zlib.
  - Datetime: This library will be used to provide a time stamp. You need to include this time stamp in every block so that the generated hash for any block will be different. For instance, if there are two blocks that are having the same data, so their hashes will be same. So, in every block, time stamp should be included so that even the blocks having the same data will be having different hash values.
  - Hashlib: This library will be used to create hash for every block. Hashlib has several built-in functions which you can use. For instance, `sha1()`, `sha224()`, `sha256()`, `sha384()`, `sha512()`, and `blake2s()`. Moreover, the reason for using `sha256()` is that it has been used in Ethereum and Bitcoin blockchains.
- Create a loop to iterate number of times user want to check the number of Hashes.
- Ask user to provide data to generate its Hash Value. The generated Hash value should consider timestamp while generating the hash value input by the user.
- You should also demonstrate that if the data input by user is same, the generated hash value will be same, in case time stamp is not added and different in case if time stamp is added.

5.4.3   Desired Program Output

The output of the program is shown in Fig. 5.2.
Output of the Program

5.4.4   Sample Code

```
1  # Lab 2 program
2
3  import datetime
4  import hashlib
5
6  print("\n")
7  print("Blockchain Systems and Communication Networks: From
       Concepts to Implementation")
8  print("===============================   Lab 2
       =======================================")
9
10 loopnmbr = int(input('Number of Hashes You Want to Check: '))
```

Fig. 5.2 Out of the program: lab2

```
11  typeofhash = int(input('The following Hash Algorithms are
        available:\n1. Sha1() '
12                          '\n2. Sha224() \n3. Sha256() \n4. Sha384()
        \n5. Blake2b() '
13                          '\n6. Blake2s() \n7. Sha3_256()'
14                          'Enter Your Choice for Generating the Hash:
        '))
15  data = str(input('Enter Data to generate the Hash Value: '))
16
17
18  class Lab2_Hashing():
19      timestamp = 0
20
21      def __init__(self,timestamp):
22          self.timestamp = datetime.datetime.now()
23
24      def hash(self):
25          string = str(self.timestamp) + data
26          string = string.encode('utf-8')
27          if typeofhash == 1:
28              h = hashlib.sha1(string)
29              print("\n")
30              print("The Computed Hash Value for SHA1")
31              return h.hexdigest()
32          if typeofhash == 2:
33              h = hashlib.sha224(string)
34              print("\n")
35              print("The Computed Hash Value for SHA224")
36              return h.hexdigest()
37          if typeofhash == 3:
```

```
38              h = hashlib.sha256(string)
39              print("\n")
40              print("The Computed Hash Value for SHA256")
41              return h.hexdigest()
42          if typeofhash == 4:
43              h = hashlib.sha384(string)
44              print("\n")
45              print("The Computed Hash Value for SHA384")
46              return h.hexdigest()
47          if typeofhash == 5:
48              h = hashlib.blake2b(string)
49              print("\n")
50              print("The Computed Hash Value for Blake2b")
51              return h.hexdigest()
52          if typeofhash == 6:
53              h = hashlib.blake2s(string)
54              print("\n")
55              print("The Computed Hash Value for Blake2s")
56              return h.hexdigest()
57          if typeofhash == 7:
58              h = hashlib.sha3_256(string)
59              print("\n")
60              print("The Computed Hash Value for SHA3_256")
61              return h.hexdigest()
62
63
64
65  def main():
66
67      for n in range(loopnmbr):
68          hashes = Lab2_Hashing(datetime.datetime.now)
69          output1 = hashes.hash()
70          print("\n")
71          print("Time Stamp: ", datetime.datetime.now())
72          print("Your Computed Hash Value: ", output1)
73
74  main()
```

5.5 Lab Implementation 3

5.5.1 Aim

The aim of this lab is to design a basic blockchain architecture using an open-source programming language such as Python.

5.5.2 Steps to Follow

You are required to perform these steps:

- First, import the Python libraries. This includes the datetime library and hashlib. If you want to use Adler32 and CRC32 then you also need to import zlib.

- Datetime: This library will be used to provide a time stamp. You need to include this time stamp in every block so that the generated hash for any block will be different. For instance, if there are two blocks that are having the same data, so their hashes will be same. So, in every block, time stamp should be included so that even the blocks having the same data will be having different hash values.
- Hashlib: This library will be used to create hash for every block. Hashlib has several built-in functions which you can use. For instance, `sha1()`, `sha224()`, `sha256()`, `sha384()`, `sha512()`, and `blake2s()`. Moreover, the reason for using `sha256()` is that it has been used in Ethereum and Bitcoin blockchains.

- Create Block Class.

  - Whenever you mine a block to blockchain, this block class comes in. Every block is an instance of Block class. This block Class should have these parameters: The specific block number, Data we want to store, Pointer to Next Block, Hash value of that specific block, Nonce, Hash of previous block, that's why blockchain is immutable, and Time Stamp.
  - All above variables are modified at the run time, e.g., every new block has different data, hash, etc.
  - The Block Class will have a `hash()` function. This `hash()` function will calculate the hash for the block. Make a string of nonce, data, previous hash, timestamp, and blocknumber (Add Them). This will give you a new Hash value every time, so every block Hash is unique and it can be identified Hash by changing Nonce, we get a completely different hash.

- Create Blockchain Class.

  - This Blockchain Class should have these parameters: the difficulty in Mining, with Respect to Target Value, maximum Nonce, and target value.
  - Here we are decreasing or increasing the target range. The less the target is, hard will be to mine block. That's how Bitcoin Controls the rate to which new blocks are mined. If diff = 0, every block gets accepted.
  - The Blockchain Class need to have two functions `add()` and `mine()`. The `add()` function is used to add a new block in the list. In the `mine()` function, just add the block at this stage. In the next lab, we will develop further this function to simulate the working of consensus algorithms.

- Call the Blockchain Class and use for the loop to generate 10 random blocks using the blockchain Class defined above.
- Print the Blocks of Blockchain. Blockchain is just like a linked List, as you will find the head, and from that head you can print all blocks. The output of your program should print the Block number, Data, and the Block Hash value.

Fig. 5.3 Output of program: lab3

5.5.3 Desired Program Output

The output of the program is shown in Fig. 5.3.
Output of the Program

5.5.4 Sample Code

```
1  # Lab 3 program
2
3  import datetime
4  import hashlib
5
6  print("\n")
```

```python
7  print("Blockchain Systems and Communication Networks: From
       Concepts to Implementation")
8  print("=============================   Lab 3
       ======================================")
9  print("===================== Blockchain Basic Structure
       ===========================")
10 print("\n")
11 loopnmbr = int(input('Number of Blocks You Want to Create: '))
12 print("\n")
13
14 # In this lab, we will be creating the basic structure of
       blockchain.
15 # There will be two main classes in this program.
16 # The first class will create blocks - BlockCls.
17 # The second class will create blockchain by linking blocks -
       BlockchainCls.
18
19
20 # First class BlockCls.
21 # The class will contain all the components of a block.
22 # This includes block number, data we want to store in a block,
       pointer to next block, hash, and time stamp. We can include other
23 # nonce, previous hash, and time stamp. We can include other
       parameters in the block as well if required.
24
25 class BlockCls:
26     blockNbr = 0   # The specific block number
27     stored_data = None   # Data we want to store
28     next = None   # Pointer to Next Block
29     hash = None   # Hash value of that specific block
30     nonce = 0
31     previous_hash = 0x0   # Hash of previous block - to make
       blockchain immutable
32     timestamp = datetime.datetime.now()   # Time stamp
33
34     def __init__(self, data):
35         self.stored_data = data
36
37 # Here we define a hash function. This hash function is SHA256,
       however, we can use any other hashing if required.
38 # See Lab # 1 and Lab # 2 for details of hash function with and
       without time stamp.
39
40     def hash(self):
41         h = hashlib.sha256()   # Here we are using SHA256 hashing.
42         h.update(
43             str(self.nonce).encode('utf-8') +
44             str(self.stored_data).encode('utf-8') +
45             str(self.previous_hash).encode('utf-8') +
46             str(self.timestamp).encode('utf-8') +
47             str(self.blockNbr).encode('utf-8')
48         )
49         return h.hexdigest()
50
51     def __str__(self):
52         return "\nBlock Hash: " + str(self.hash()) + "\nBlock
       Number: " + str(self.blockNbr) + "\nBlock Data: " + str(
53             self.stored_data) + "\n---------------"
54
```

```
55  # Second Class BlockchainCls.
56  class BlockchainCls:
57      mining_difficulty = 20   # Mining difficult is defined over
        here.
58      maximum_Nonce = 2 ** 32   # This is the maximum nonce number.
        It is 2^32, max number you can store in 32
59      target = 2 ** (256 - mining_difficulty)   # Here Target is
        2^(256-the difficulty) (e.g. if diff=20 it is 2^236)
60      # Here we are decreasing or increasing the target range. The
        less the target is, hard will be to mine block
61      # This is how Bitcoin Controls the rate to which new blocks
        are mined
62      # If diff = 0, every block gets accepted,
63
64      block = BlockCls("Genesis Block")   # This is the very first
        block - Genesis Block
65      dummy = head = block
66
67      def add(self, block):
68          # This is used to add a new block into the list
69          block.previous_hash = self.block.hash()
70          block.blockNbr = self.block.blockNbr + 1
71
72          self.block.next = block
73          self.block = self.block.next
74
75  # This is the mining function. If we want to include any consensus
        algorithm,
76  # we need to program the logic of consensus algorithm in this
        function.
77      def mine(self, block):
78          for n in range(self.maximum_Nonce):
79              self.add(block)
80              break
81
82  blockchain = BlockchainCls()
83
84
85  for n in range(loopnmbr):
86      print("Block Number: " + str(n+1))
87      inserted_data = str(input('Enter Data to include in the block
        number: ' ))
88      blockchain.mine(BlockCls(inserted_data))
89
90  # Printing Blocks of Blockchain
91
92  while blockchain.head != None:
93      print(blockchain.head)
94      blockchain.head = blockchain.head.next
```

5.6   Lab Implementation 4

5.6.1   Aim

The aim of this lab is to check the validity of the blockchain.

5.6.2   Steps to Follow

Follow the steps mentioned below:

- Check Hash Values of all linked blocks, and make sure every block is linked with next.
- Calculate Hash independently and compare it with stored hash to ensure there is no tampering in the data.

Tips:

- Create class `Block` and class `Blockchain`, as mentioned in Lab1.
- Create class `ChainValidation`. In this class, create two functions: `head_check()` and `integrity_check()`. The `head_check()` function should see and compare the hashes of each block and see if the previous hash value mentioned in this block header matches with the actual hash value of the previous block. If the hashes are not matching, print the message that blockchain not linked properly, else print that Blocks are linked properly. The `integrity_check()` function calculate hash independently and compare it with stored hash to ensure that there is not tampering in the data.

5.7   Lab Implementation 5

5.7.1   Aim

The aim of this lab is to implement Proof-of-Work (PoW) consensus algorithm of the blockchain.

5.7.2   Steps to Follow

Follow the steps mentioned below:

- Refer to Lab # 3, there you provided different parameter values in Blockchain class, such as difficulty in mining, maxNonce, and target value.

- Using the above values, incorporate a function `mine()` in Blockchain class which will implement the logic of Proof-of-Work (PoW) consensus algorithm. For illustration purpose, see out below, where when we increase the difficulty, the number of hashes required to mine the block increases. In the `mine()` function, you just need to implement this logic: If the value of Hash is less than target, the block is not mined, else increment the nonce and the block is mined successfully. Then print the number of hashes required to mine the block.

5.8 Hands-On Exercise

5.8.1 Exploring Real Blockchain: Bitcoin

Let's explore Bitcoin blockchain and understand various items in a block. Figure 5.4 shows an image taken from the website https://www.blockchain.com/.

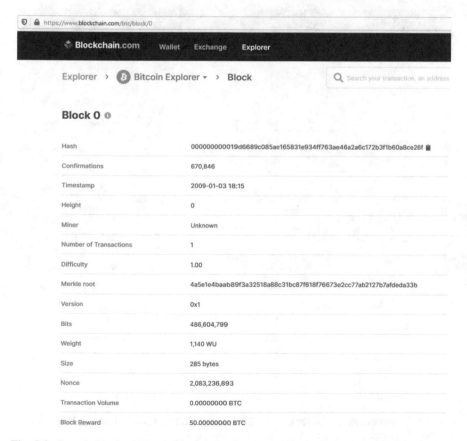

Fig. 5.4 Genesis block of Bitcoin blockchain showing height, hash, previous block hash, time, difficulty, bits, number of transactions, and output total

1. Visit the website: https://www.blockchain.com/explorer
2. Explore this website and read the items, click and explore these items.
3. Let's start from the genesis block where it all began: 0 in the search box and choose Bitcoin from drop down list.

5.8.2 Exploring Real Blockchain: Ethereum

Let's explore Ethereum blockchain and understand various items in a block. Figure 5.5 shows an image taken from the website https://www.blockchain.com/.

1. Visit the website: https://www.blockchain.com/explorer
2. Explore this website and read the items, click and explore these items.
3. Let's start from the block height 0 where it all began: 0 in the search box and choose Ethereum from drop down list.

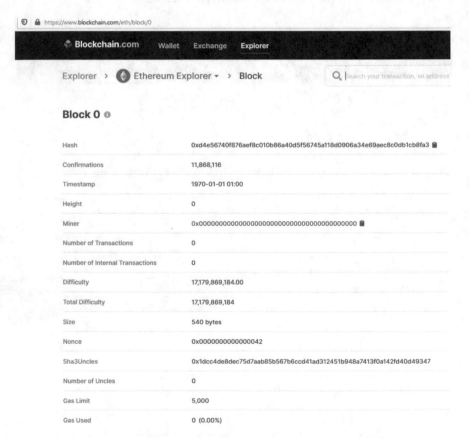

Fig. 5.5 Ethereum block at height 0 shows hash value, confirmations, timestamp, height, miner, number of transactions, and number of internal transactions

5.8.3   Exploring Real Blockchain: Bitcoin Cash: Fork of Bitcoin

Let's explore Bitcoin Cash (fork of Bitcoin blockchain) and understand various items in a block. Figure 5.6 shows an image taken from the website https://www.blockchain.com/.

1. Visit the website: https://www.blockchain.com/explorer
2. Explore this website and read the items, click and explore these items.
3. Let's start from the genesis block where it all began: 0 in the search box and choose Bitcoin Cash from drop down list.

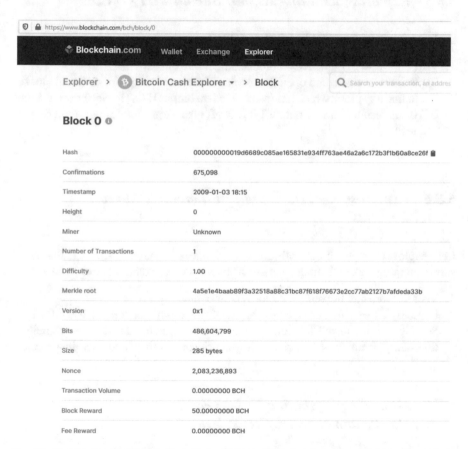

Fig. 5.6 Genesis block of Bitcoin Cash (fork of Bitcoin blockchain) showing height, hash, previous block hash, time, difficulty, bits, number of transactions, and output total

5.8.4 Exploring Real Blockchain: Bitcoin Blocks Linkage

Let's explore how Bitcoin blockchain's blocks are linked together. Figure 5.6 shows an image taken from the website https://www.blockchain.com/.

1. Visit the website: https://www.blockchain.com/explorer
2. Explore this website and open these blocks of Bitcoin blockchain: 100000, 100001 and 100002.
3. See how previous block is referenced in the current block.

5.8.5 Exploring Real Blockchain: Bitcoin's UTXO Concept

Let's explore how Bitcoin blockchain's UTXO work.

1. Visit the website: https://www.blockchain.com/explorer
2. Explore this website and open the block of Bitcoin blockchain: different blocks.
3. Try to find any block which has resulted in two output UTXOs. See how one input UTXO is resulted in two output UTXOs. It also shows total input, total output, and fees.

5.8.6 Exploring Real Blockchain: Ethereum's Block Contents

Let's explore Ethereum blockchain and understand various items in a block. Figure 5.7 shows an image taken from the website https://www.blockchain.com/.

1. Visit the website: https://www.blockchain.com/explorer
2. Explore this website and open the block of Ethereum blockchain: 1000000.
3. See Fig. 5.7, it shows Ethereum's block content showing TxHash, block height, timestamp, from, to, value, gas limit, gas price, and other fields within a block.

5.8.7 How Many Byzantine Nodes (Faulty Nodes) a Blockchain Network Can Tolerate?

Example: Suppose there are $N = 20$ consensus nodes in a blockchain network. This blockchain network can reach to consensus—$\lfloor \frac{N-1}{5} \rfloor$ to $\lfloor \frac{N-1}{2} \rfloor$ toleration level—in the presence of Byzantine nodes. How many Byzantine nodes (faulty nodes) this blockchain network can tolerate?

Fig. 5.7 Ethereum's block content showing TxHash, block height, timestamp, from, to, value, gas limit, gas price, and other contents

Solution:

Let's take $N = 20$. This blockchain network can tolerate 3–9 faulty nodes.

$$\lfloor \frac{N-1}{5} \rfloor = \lfloor \frac{20-1}{5} \rfloor = \frac{19}{5} = 3.8 = 3, \tag{5.1}$$

$$\lfloor \frac{N-1}{2} \rfloor = \lfloor \frac{20-1}{2} \rfloor = \frac{19}{2} = 9.5 = 9, \tag{5.2}$$

5.8.8   How to Find the Size of Ethereum Blockchain?

Let's explore how to find the size of Ethereum blockchain?
  Visit these websites for finding the current size of Ethereum blockchain:

- https://etherscan.io/chartsync/chaindefault
- https://blockchair.com/ethereum/charts/blockchain-size.

5.8.9   How to Find the Transaction Handling Capacity of Blockchain?

Let's explore how to find the transaction handling capacity of blockchain?
  Transaction rate per second can be found on this website:
https://www.blockchain.com/charts/transactions-per-second.

5.8.10   How to Find Tx Throughput and Block Confirmation Time

Suppose a blockchain network has a block size of 4 MBytes, transaction size of
100 Bytes, Block time is 400 s and it requires to wait for 20 blocks to confirm any
new block. Calculate average block confirmation time and transaction per second
(throughput) of this blockchain network?

$$Transaction\ Throughput = \frac{4\ (MBytes)}{100\ (Bytes) \times 400\ (s)}, \tag{5.3}$$

$$Transaction\ Throughput = \frac{4 \times 10^6}{100 \times 400}, \tag{5.4}$$

$$Transaction\ Throughput = 100\frac{Tx}{seconds}, \tag{5.5}$$

$$Average\ Confirmation\ Time = 20 \times 400\ s = 8000\ s, \tag{5.6}$$

The Transaction Throughput is 100 Tx/s and Average Confirmation Time is 8000 s.

5.8.11   How to Find Wining Probability in PoW Consensus

Suppose a blockchain network with seven participating nodes ($N_1, N_2, \ldots, N_7$),
having computing power (hash rate) values ($\varphi_1, \varphi_2, \varphi_3, \ldots, \varphi_7$). The computation power (hash rates) are: $\varphi_1 = 18, \varphi_2 = 14, \varphi_3 = 15, \varphi_4 = 100, \varphi_5 = 30, \varphi_6 =$

16, $\varphi_7 = 150$. Calculate the probability of wining the puzzle by each participating node?

The winning probability P_w of each participating node can be calculated as:

$$P_{w_i} = \frac{\varphi_i}{\sum_{j=1}^{N} \varphi_j}, \qquad (5.7)$$

where i is the participating blockchain node, N is the total number of nodes, and φ is the computation power of ith node.

The wining probability of participating node N_1 can be calculated as:

$$P_{w_1} = \frac{\varphi_1}{\sum_{j=1}^{7} \varphi_j}, \qquad (5.8)$$

$$P_{w_1} = \frac{18}{18 + 14 + 15 + 100 + 30 + 16 + 150}, \qquad (5.9)$$

$$P_{w_1} = 0.052, \qquad (5.10)$$

Similarly, the wining probability of participating node N_7 can be calculated as:

$$P_{w_7} = \frac{\varphi_7}{\sum_{j=1}^{7} \varphi_j}, \qquad (5.11)$$

$$P_{w_7} = \frac{150}{18 + 14 + 15 + 100 + 30 + 16 + 150}, \qquad (5.12)$$

$$P_{w_7} = 0.43, \qquad (5.13)$$

5.9 Summary

This chapter provided implementation details of blockchain system along with hands-on exercises to interact with real blockchain system. First, two mini projects were discussed which helps the reader to critically analyze distributed ledger technology and then implement distributed ledger technology and perform it's security analysis using Microsoft's STRIDE threat analysis. Lab implementations were then discussed in detail where aim, steps to follow, desired program output, and the sample code were provided to help understand the readers the implementation process of blockchain systems. Finally, few hands-on exercises were provided about some basic concepts of blockchain.

Part III
Blockchain Systems and Communication Networks

This part of the book talks about blockchain systems and communication networks and two chapters are included. The first chapter (chapter six) discusses cognitive radio networks and blockchain. The second chapter (chapter seven) talks about communication networks and blockchain in general covering various communication networks such as Wi-Fi, cellular networks, cloud computing, Internet of Things, software defined network, and smart energy networks.

Chapter 6
Cognitive Radio Networks and Blockchain

The history of communication dates back to the history of mankind. In the early ages, mankind used different methods of communication ranging from paintings in a cave to the use of wooden blocks. Writing was another way of communication for which different languages were developed. People also used different forms of signals to deliver their messages over a distance. These signals were in the form of smoke or using different flag colors to highlight any message. Telegraph was another method in which different symbolic codes were used to deliver messages over long distances. Optical telegraphy and electric telegraphy are the examples of such systems. With the advent of electric systems, researchers and scientists used them for communication. In this context, electric telephone (landline telephone) was invented back in the late 1870s. These inventions and progress did not stop here. Radio and television were invented and used for communication in the early 1900s. The concept of mobile telephone was proposed in the late 1940s at Bell Labs which resulted in the first mobile network, Nordic Mobile Telephone, in 1981. The Global System for Mobile Communication (GSM) standard was developed in 1991.

In parallel, there were also significant developments in Computers and Internet. In 1949, Claude Shannon, mathematically proves the Nyquist-Shannon Sample Theorem. An exponential rise in inventions can be seen in late 1980 in the domain of the Internet. In 2020, when we see around us, we are now talking about Tactile Internet, remote surgery, Internet of Things, and smart grid communication. All areas of human life are fully integrated and dependent on these communication networks. In COVID-19 global pandemic, schools and universities were shut down, countries are fully locked down, flights are stopped, and even people are not allowed to move from their houses except for buying essential goods or for urgent medical needs. Business and shops are closed. Thanks to the advancements in communication networks that now as an alternative, schools moved online, universities are delivering their modules online, shopping can be carried out online, and remote counseling from medical experts and general practitioners can now be possible remotely.

© Springer Nature Switzerland AG 2021
M. H. Rehmani, *Blockchain Systems and Communication Networks: From Concepts to Implementation*, Textbooks in Telecommunication Engineering,
https://doi.org/10.1007/978-3-030-71788-9_6

Table 6.1 Wireless- and Wire-based communication systems with their data rates

Wireless and wired communication systems	Supporting data rates
IEEE 802.3x	1 Gbps
SONET/SDH (Fibre Optic)	10 Gbps
ADSL	1–8 Mbps
HDSL	2 Mpbs
VDSL	15–100 Mbps
PLC	14–200 Mbps
Wi-Fi IEEE 802.11x	2–600 Mbps
ZigBee	250 kbps
Cellular 2G	14.4 kbps
Cellular 2.5G	144 kbps
Cellular 3G	2 Mbps
cellular 3.5G	14 Mbps
Celular 4G	100 Mbps
WDM (Fiber Optic)	40 Gbps
Z-wave	40 kbps

6.1 Wired and Wireless Communication Systems

Modern communication systems can be classified into wired-based and wireless-based. Table 6.1 shows the list of wired and wireless communication systems along with their supporting data rates.[1] In wired-based communication systems, wire is used as a communication medium to transfer data from one place to another. These wire-based communication systems generally provide higher bandwidth and transfer a huge amount of data very quickly.

In comparison with wire-based communication systems, wireless-based communication systems do not rely on the physical wire. Instead, wireless signals are used to transfer data from one place to another. Like other natural resources, such as petrol, gas, and minerals, wireless radio spectrum is also a natural resource. A small chunk of wireless radio spectrum band cost millions of dollars. Wireless radio spectrum band can start from few kilohertz (KHz) and goes till megahertz (MHz), gigahertz (GHz), and terahertz (THz). These frequencies have their own unique features and can be used in different communication network applications. Figure 6.1 shows the behavior of wireless radio spectrum from lower frequencies to higher frequencies in terms of "coverage area", "transmit power", "power consumption", and "bandwidth". These frequency bands are regulated by international and national government agencies and communication networks are established by buying the spectrum band from the government for normal operations of these networks.

---

[1]More details about wired and wireless communication systems and their associated data rates can be found in [49, 58].

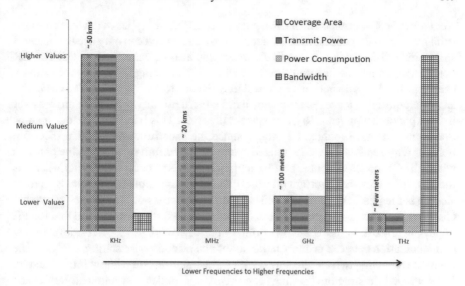

Fig. 6.1 Behavior of wireless radio spectrum from lower frequencies to higher frequencies in terms of "coverage area", "transmit power", "power consumption", and "bandwidth"

This wireless radio spectrum can also be classified as licensed spectrum and unlicensed spectrum. The licensed spectrum is managed by the respective governments and can only be used by buying it. Telecom operators buy these licensed bands and then provide communication services to their customers. In contrast, unlicensed spectrum is the one which can be used free of charge by anyone at any time. This unlicensed spectrum can also be called as Industrial, Scientific, and Medical (ISM) band.

The United States Federal Communication Commission (FCC) made a study and observed that utilization of this wireless radio spectrum varies according to the time and geographic location. They showed that the wireless radio spectrum is underutilized, and lot of licensed radio spectrum is not being utilized. The primary reason for this is the static allocation of the spectrum by the regulatory bodies. In order to efficiently utilize the wireless radio spectrum, researchers and scientists proposed to use this wireless radio spectrum dynamically. Thus, a new paradigm was introduced known as Dynamic Spectrum Access (DSA).

6.2   Dynamic Spectrum Access (DSA)

In DSA, first, the spectrum is monitored and then if not utilized by the licensed users, that portion of the wireless radio spectrum can be used by unlicensed users or it can be used for lease purpose. Cognitive Radio (CR) is the technology that enables this DSA paradigm to efficiently utilize this unused wireless radio spectrum. A Cognitive Radio

Network (CRN) is composed of CR devices. In a CRN, two types of devices co-exist: Primary Radio (PR) nodes, which are the licensed users and can only use the licensed spectrum, and Secondary Users (SU), which can also be called CR nodes. The CR nodes can only use the licensed spectrum when it is not being utilized by the primary users and CR nodes do not cause harmful interference to the PR nodes. The CR nodes are also required to leave the frequency band when a primary radio node arrives over it. This phenomenon leads to CR spectrum lifecycle. This lifecycle has four phases. Spectrum sensing, in which CR nodes sense the spectrum and identify spectrum holes or white spaces, i.e., non-utilized spectrum. Spectrum selection is the phase in which the CR nodes select the best frequency band based upon some criteria such as the time for which the spectrum is unoccupied or the available bandwidth. Spectrum sharing is the phase in which CR nodes co-ordinate access of the channel with other CR users. In spectrum mobility, CR nodes release the frequency band so that the PR nodes can reuse it. Figure 6.2 shows primary radio network (on the left side) showing communication between primary radio nodes and primary base station. CRN (on the right side) showing two architectures: with infrastructure, and without infrastructure. In "without infrastructure" setting, cognitive radio nodes communicate with each other and perform spectrum lifecycle on their own. In "with infrastructure" setting, CR nodes communicate with each other through cognitive base station and it's the cognitive base station which performs spectrum lifecycle. Figure 6.3 shows an ad hoc CRN is formed in the presence of primary radio network in which CR nodes communicate with each other using different channels owned by the primary network when these channels are unoccupied by the primary radio nodes.

The spectrum scarcity and underutilization are due to two main reasons. The first reason is the nature of wireless radio spectrum, i.e., how the frequency bands behaves, and it is governed by the laws of Physics. The second reason is the way through which the assignment is done by the regulatory bodies (static spectrum assignment in this case). Both these factors when accumulated together with the demand for new wireless services result in spectrum scarcity or underutilization.

6.3 Blockchain and Spectrum Management

Blockchain can help to establish trust among different organizations managing and involving in the spectrum regulation including spectrum sharing and trading market. It can also help to store data related to the complete life cycle of spectrum management, i.e., from spectrum regulation, spectrum management to spectrum sharing, and spectrum trading. This technology will not only help to make the system more transparent but also auditable and immutable. When considering wireless radio spectrum as an asset, it can be tracked from the time it is assigned by the regulator and then divided into chunks of usable frequency bands by the mobile network operator and then assigned to the Base Transceiver Station (BTS), and then its assignment to the incumbent users and then it's trading to/by the SU. More advanced features of blockchain, such as smart contracts can also revolutionize the spectrum trading

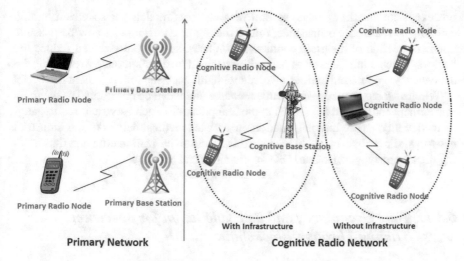

Fig. 6.2 Primary radio network (on left side) showing communication between primary radio nodes and primary base station. CRN (on right side) showing two architectures: with infrastructure, and without infrastructure. In "without infrastructure" setting, CR nodes communicate with each other and perform spectrum lifecycle on their own. In "with infrastructure" setting, CR nodes communicate with each other through cognitive base station and it's the cognitive base station which performs spectrum lifecycle

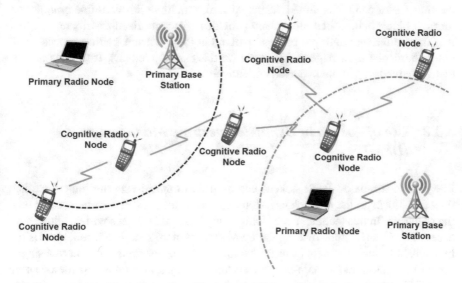

Fig. 6.3 An ad hoc CRN is formed in the presence of primary radio network in which CR nodes communicate with each other using different channels owned by the primary network when these channels are unoccupied by the primary radio nodes

process. With smart contracts, several new business models for spectrum trading can now be realized. For instance, future trading of spectrum can now be possible upon completion of a certain condition. Similarly, smart contracts can reduce the business process time as well as make the system more autonomous by removing the dependency on different entities involved with this spectrum management life cycle.

Wireless radio spectrum is a scarce resource and it can be considered as the main impediment for beyond 5G (B5G) and 6G applications such as virtual reality, augmented reality, tactile Internet, and remote assisted medical networks and industrial networks. Blockchain can help to make this spectrum utilization efficient, thus helping directly to support B5G and 6G applications and services.

6.3.1   Time Granularity and its Exploitation for Spectrum Trading Through Blockchain

Consider the availability of spectrum in terms of time. Spectrum can be traded not only for the duration of months, weeks, and days but it can also be sold on short periods of time, such as hours, minutes, and even seconds. This new paradigm of spectrum trading for very short periods of time will enhance spectrum utilization as well as it will improve revenue generation. Figure 6.4 shows this concept that how the time factor can be exploited during spectrum trading with very fine granularity through blockchain. Since blockchain provides a decentralized distributed ledger, it is imperative that multiple parties involved in the spectrum trading process will not be burdened to manage the database (in this case the ledger). It means that the distributed ledger will be managed collectively.

6.3.2   Use of Tokens in Dynamic Spectrum Management (DSM)

In a blockchain-based DSM, tokens can be used to incentivize the miners so that they add blocks to the blockchain. Moreover, these tokens can also be used to buy the spectrum. In this manner, nodes which only have these tokens will be allowed to access the channels, thus making the DSM system more secure and reliable. Using blockchain in such a system ensures that the deployed system is resilient to a single point of failure. In blockchain-based spectrum trading, spectrum is considered as an asset. This asset can be traded and transferred from one person to another with the help of tokens. One may buy these tokens by using fiat currency or cryptocurrency or by providing any service such as mining or spectrum sensing or spectrum sharing.

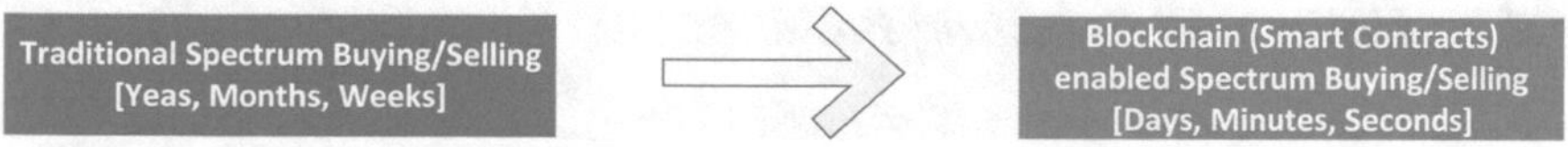

Traditional Spectrum Buying/Selling [Yeas, Months, Weeks]	⟹	Blockchain (Smart Contracts) enabled Spectrum Buying/Selling [Days, Minutes, Seconds]

Fig. 6.4 Spectrum trading in traditional and blockchain enabled based systems

6.4 Usage of Blockchain Technology from the Spectrum Licensing Perspective

From the perspective of licensing, white spaces can be considered in licensed band, shared licensed band, and unlicensed band.

6.4.1 Licensed Spectrum Band

In licensed band, only the licensed user can get access to the band, and users who do not have a legitimate license are not allowed to use licensed frequency bands. The users of licensed band want to share their spectrum to generate more revenue or to get incentive from the regulator in terms of paying less fees for the license. The CR users who want to use these licensed band also consider some factors when accessing and getting involved in the trading of licensed spectrum. For instance, CR users get high chances to get access to the spectrum band as a major portion of the spectrum is licensed. Moreover, licensed band offers very good channel characteristics such as high bandwidth or long-distance propagation, both factors motivate CR users to use licensed spectrum band for their usage.

6.4.2 Shared Licensed Spectrum Band

In shared licensed band paradigm, spectrum is shared among the parties which hold the licensed for a portion of the spectrum. In this case, multiple licensed holder parties can share the same portion of the spectrum but with certain rules mentioned in the shared licensed agreement. In this shared licensed spectrum paradigm, users get good QoS as compared to unlicensed spectrum, less interference to other users, as there are limited parties which are allowed to access the same spectrum with certain rules, and finally, access to the spectrum can be guaranteed to a higher extent.

6.4.3 Unlicensed Spectrum Band

In unlicensed spectrum paradigm, users can access the spectrum without any charges and permission, thus allowing anyone to access any portion of the unlicensed spectrum freely. The users in unlicensed spectrum band want to coordinate with other users to guarantee some QoS, to limit harmful interference, and to reduce contention in accessing the channels. Blockchain can play a crucial role in these spectrum sharing paradigms from the licensed perspective.

6.5 Blockchain Enabled Cognitive Radio Network and Collision-Free Communication

6.5.1 Collision-Free Communication

Collision-free communication (CFC) is desirable in wireless and wired networks and this has been a topic of research for several decades. Numerous medium access protocols were proposed and standardized. The role of those medium access protocols is to allow devices to access the medium (resource) without making a collision with other devices. Figure 6.5 shows a classification of medium access protocols used in different communication networks. Avoiding collision is necessary; otherwise, packets will be lost, re-transmission will be required and thus result in energy and time consumption.

In cellular networks, such as 2G, 3G, and 4G, various scheduling mechanism are applied (TDMA, CDMA, FDMA) to assign resources to mobile terminals. This is typically done when a mobile station (MS) contact the base transceiver station (BTS) and request to allocate a resource in order to perform uplink transmission. In response, BTS assign a channel to the MS for its uplink transmission. In emerging

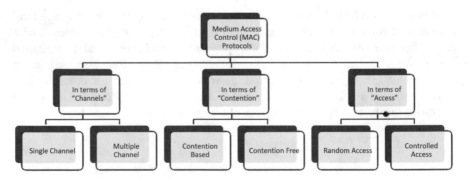

Fig. 6.5 Medium access control (MAC) protocols for communication networks. These MAC protocols are classified in terms of "channels", "contention", and "access"

cellular band technologies, such as machine type (MT) communication, the amount of traffic is less, thus scheduling mechanisms are not suitable in such scenarios. These scheduling mechanisms can also be termed grant-based mechanism. In contrast, grant-free scheduling mechanisms are required where the BTS (central entity) do not allocate the resource for the uplink transmission of MS. In this context, blockchain can play its role in achieving a consensus among multiple nodes to decide which radio resource needs to be used by which mobile station at a specific time. This can be done in a decentralized manner and thus reducing collision among the participating nodes.

6.5.2 Blockchain-Enabled Cognitive Radio Network and CFC

In a blockchain-enabled CRN, a group of nearby nodes is established. These groups of nodes are responsible for allocation of radio resource within the group. The nodes in each group within the CRN share the set of resources they intend to use. This subset of resources will first get validated among the group members and once resources are assigned without any conflict, such resource allocation be broadcasted and add to the blockchain. In this manner, efficient resource allocation will be carried out within the CRN. All the nodes in the network will be aware of which resources are allocated to which nodes and thus reducing the chances of collision within the CRN. Here in this context, the decentralized property of blockchain plays an important role to reach consensus among the nodes without the presence of a centralized node. Second, the whole network nodes will be aware of assigned resources among the CR nodes, as blockchain provides the copy of the ledger (in this case, the copy of ledger will contain allocated resources to devices) to every node and every CR node will have a global view of allocated network resources.

Figure 6.6 shows the scenario where CRN is organized into two groups for resource allocation. That is, Group # 1 and Group # 2. Nodes in each of these groups are responsible for radio resource allocation within the group. The following steps will be carried out to avoid collision-free communication in CRNs.

1. Step # 1: Local consensus is reached to allocate resources in each group.
2. Step # 2: Broadcast this information to all the CR nodes in the network.
3. Step # 3: CR nodes will not use these resources.
4. Step # 4: Results in collision avoidance in channel hopping sequence-based CRN.

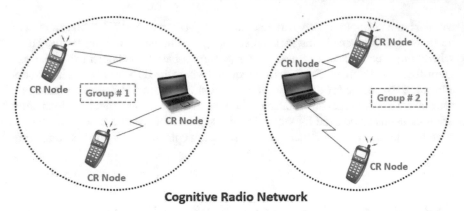

Fig. 6.6 A CRN is organized into two groups, i.e., Group # 1 and Group # 2. Nodes in each of these groups are responsible for radio resource allocation within the group

6.6   Medium Access by CR Nodes as an Auction

In CRNs, finding spectrum opportunities is crucial as it will help to increase spectrum utilization of under-utilized spectrum. Spectrum access (medium access) layer of CRNs helps CR devices to access the spectrum. In a centralized CRN, a central spectrum broker can suggest a particular spectrum band to a CR device, however, in a distributed and decentralized CRN, CR devices need to perform spectrum sensing themselves along with deciding which spectrum to access at a particular time given that the spectrum is not being utilized by the PR nodes.

In order to increase the performance of medium access protocols in CRNs, auctions mechanism can be integrated into MAC protocols, or in simple words, medium access by the CR nodes can be treated as an auction. In this context, auction bids can be launched by the nodes which have identified spectrum as not utilized by themselves (if PR nodes are not using it). Moreover, the PR nodes can also launch the bidding of their spectrum and interested CR nodes can involve in the bidding process to utilize that spectrum. It may also be possible that a particular spectrum band is available, i.e., no PR activity found on that channel and now multiple CR nodes want to access the channel. Now since multiple CR nodes are accessing the same channel, contention will occur, and it will result in a collision and CR nodes may lose some good opportunities for spectrum utilization. Here in this context, auction theory can be used to help CR nodes to efficiently access the spectrum.

Auction theory has been developed for decades and has been applied to many application areas. In telecommunication networks, auction theory has been applied as well. These auction mechanisms can be classified as single auction strategies and repeated auctions. Double auction is another well know auction applied to telecom networks. Each auction category has its own advantages and drawbacks. These auction strategies together with blockchain systems can help a lot to improve the overall performance of communication networks.

6.7   Advantages of Using Blockchain Technology in Dynamic Spectrum Management (DSM)

Blockchain brings several advantages to the dynamic spectrum management paradigm. Blockchain-based dynamic spectrum management system can be applied to different telecommunication systems such as Wi-Fi (ISM band), TV White Space (TVWS), satellite communication, D2D networks, and Citizen Band Radio Service (CBRS). The mining process and the associated rewards (incentives) for successful mining can further enhance spectrum usage. More precisely, the user will participate in mining process, blocks will be added to the ledger—making ledger up-to-date, and further incentivizing users to mine. On the other side, users will share more and more of their spectrum to get more rewards, thus making the spectrum available to other nodes in the network. This will also enable nodes to get the lease of the spectrum in future. Below we discuss some features of blockchain that can help in dynamic spectrum management.

6.7.1   Lack of Central Entity

Blockchain through its decentralization feature can remove the reliance of buying and selling spectrum without going to the central entity or spectrum regulator. This can enhance the spectrum assigning process by making the buying and selling more faster than getting permission from a centralized regulator (which may not be available in bank holidays or on weekends).

6.7.2   Immutability

This feature of blockchain can help in keeping the spectrum management-related records safe and secure. Moreover, the records cannot be tampered easily in a blockchain system designed for spectrum management. This feature makes blockchain-based dynamic spectrum management system more transparent and one can easily audit the previous spectrum-related transactions.

6.7.3   Availability

This feature of blockchain makes the dynamic spectrum management distributed ledger more available as several copies of the ledger co-exist and it will be extremely difficult for the attacker to make modifications in the ledger. Additionally, there will be no need to contact regional or national spectrum regulator every time when evaluating white spaces and determining spectrum opportunities. Consider the fine granularity of spectrum opportunities available (in days, hours, and minutes) and then make an auction of those available opportunities very quickly can only be possible with the help of blockchain-based dynamic spectrum management.

6.7.4 DoS Resilient

Blockchain-based dynamic spectrum management system makes the system denial of service attack resilient as the copy of the ledger is available at multiple nodes and a compromise to a single node or few nodes will not hinder the activities of the blockchain system.

6.7.5 Non-repudiation

This non-repudiation feature of blockchain makes the dynamic spectrum management more transparent and auditable since once the transactions related to dynamic system management is recorded into the ledger, a node cannot disown it, thus a completely transparent record of transaction can be managed and maintained.

6.7.6 Smart Contract Integration

Blockchain brings smart contract feature which will bring new business models and rule-based spectrum management and assignment policies.

6.8 Spectrum Patrolling Through Blockchain

In cellular networks, mobile network operators (MNOs) can use the wireless radio spectrum opportunistically of the other MNOs or primary networks or secondary networks in order to meet their radio spectrum requirements. However, this spectrum utilization by the MNOs can only be possible if the MNOs have spectrum utilization information; otherwise, the MNO may cause harmful interference to the licensed users or SUs. In order to deal with this, MNO can have two possible ways: either perform the spectrum sensing itself or ask other users in the system to provide spectrum sensing-related information. In a systematic way, MNO can also get in touch with other spectrum sensing service providers and utilize the unoccupied spectrum. However, in this case, the MNO needs to pay the spectrum sensing service provider payment for using this service. It may be possible that the users provide their spectrum

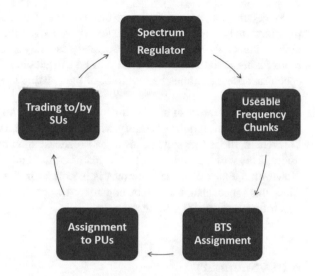

Fig. 6.7 Tracking of wireless radio spectrum from spectrum regulator to spectrum users

sensing results to the MNO and get a reward or incentive for providing such a service. This will be led to the crowdsource spectrum sensing regime. The users in this crowdsource spectrum sensing paradigm can provide their spectrum sensing results to the MNO and then the MNO aggregate the spectrum sensing results and make an appropriate spectrum decision. All these interactions between the users and the MNO need to be documented and appropriate transaction need to be taken place among these multiple parties, thus resulting in spectrum sensing-related information trading.

Spectrum policymakers and regulators need to monitor the utilization of spectrum and to take appropriate actions for any spectrum utilization breach. This can be done by using crowdsourced spectrum sensing and dedicated spectrum monitors deployed in a particular region. This type of activity is referred to as "spectrum patrolling" in the literature (see further reading section for specific reference). Figure 6.7 shows tracking of wireless radio spectrum from spectrum regulator to spectrum users.

6.9 Issues and Challenges When Deploying Blockchain to Dynamic Spectrum Management

When thinking about deploying blockchain-based dynamic spectrum management system, it is important to think about who will manage this blockchain? Will the spectrum broker own this ledger? How the nodes join and leave the blockchain network? How the consensus will be reached among the blockchain nodes? Which consensus protocol will be more feasible? Which blockchain system, i.e., public or consortium or private is appropriate?

Since the wireless radio spectrum will be acquired by the mobile devices, therefore, they need to communicate as well over the wireless channel. As we already know, wireless channel has already capacity issues, thus it is not clear how blockchain-related communication will be carried out in such a scenario? Will new spectrum access methods be required, or the existing spectrum access methods will suffice?

One unique feature of blockchain system is that the nodes keep their local copy of the ledger and this requires memory. Moreover, each time when a block is added to the blockchain, the ledger state needs to be broadcast and update among the blockchain nodes. This will also incur cost in terms of communication and memory overhead. However, the lack of these resources is important in the context of mobile devices. Thus, this aspect also needs to be considered when thinking about blockchain-based spectrum management systems.

6.10 Summary

In this chapter, we discussed the history of communication and then we highlighted different modes of communication, i.e., wired and wireless communication systems. Afterward, we focused on dynamic spectrum management and the use of blockchain technology for dynamic spectrum management from different perspectives. We also discussed the motivation behind using CRNs. We also discussed how blockchain can exploit different paradigms of white space. We then discussed the advantages of using blockchain technology for DSM and we concluded the chapter by discussing issues, and challenges associated with the usage of blockchain technology in CRNs.

6.11 Future Research Directions

One possible open research area is the dynamic channel behavior of spectrum in CRNs. More precisely, channel dynamics can be due to fading channel characteristics and due to PR activity. It is important to consider both channel dynamics when designing future auction MAC protocols for blockchain-enabled CRNs.

6.12 Further Reading

To read further on wireless and wired communication standards, [43] is a wonderful resource. History of IEEE 802 standard can be found in [94]. A discussion on CR can be found in [48, 49]. Discussion about the use of distributed ledger technology for wireless networking can be found in [57]. An interesting discussion on drone is present in [46]. Reinforcement learning and CRNs are discussed in [80]. Cybersecurity framework for vehicular network is discussed in [81]. Network coding for CRN is presented in [64]. White space [4], and CRNs are discussed in [76].

6.12.1 Blockchain and Spectrum Management

Many research articles [53, 82, 93] are present on the applicability of blockchain for spectrum management, and the use of blockchain technology for CRNs. Spectrum trading using blockchain and using smart contracts for spectrum sensing can be found in [17, 82].

Problems

6.1 What is the motivation behind cognitive radio networks?

6.2 Explain the behavior of wireless radio spectrum when moving towards lower frequencies to higher frequencies?

6.3 Assume a cognitive radio network scenario in which spectrum sensing is carried out as a service. There are N CR users who perform sensing. Let's assume a amount to be paid for sensing by each CR user (without considering time granularity). Calculate how much gas will be paid in an Ethereum network?

6.4 Explain dynamic spectrum access.

6.5 Why there is a need for spectrum management through blockchain?

6.6 How collision free communication can be achieved through blockchain enabled cognitive radio networks?

6.7 Explain in your own words that how blockchain can facilitate spectrum auction in cognitive radio networks.

6.8 Define spectrum patrolling?

6.9 What are issues and challenges when deploying blockchain for dynamic spectrum management?

Chapter 7
Communication Networks and Blockchain

When thinking about applying blockchain systems for telecommunication networks, we need to think about what benefits this blockchain technology will bring to the telecommunication system under consideration? How this blockchain technology when applied to telecommunication network is superior to traditional database solutions? What cost needs to be incurred when deploying blockchain technology to telecommunication networks? In this chapter, we discuss the answer to these questions and highlight the potential application of blockchain technology in various communication networks. Table 7.1 shows various functionalities that can be achieved in wireless networks through blockchain.

A Distributed Ledger (DL) is basically a ledger that keeps digital data replicated, synchronized and shared over several machines (nodes) in a geographical distant location, without administering these machines centrally. This DL is responsible for providing a common view to the nodes geographically placed in distant locations. This common view can also be referred to as "consensus". This consensus is not limited to digital data in the form of transaction record, instead, it can be the configuration of multiple nodes or it can be a strategy to store data in Storage Area Network (SAN) or it can be a policy to adopt by multiple nodes to achieve a certain task in a network. These features of DLT, reaching consensus in a geographically distributed nodes without a centralized entity, resembles different communication networks (cf. Table 7.2 for similarities between different communication networks and DLT). Thus, one may think about potential applicability of DLT to these communication networks for different applications. For instance, a Mobile Ad Hoc Network (MANET) has no central entity and nodes are mobile in nature and free to join and leave the network at any time. Another example is a multi-hop cognitive radio network (cf. Fig 6.3 and Chap. 6 in general for more details) in which cognitive radio nodes are responsible for different spectrum-related functionalities such as spectrum sensing, spectrum selection, spectrum sharing, and spectrum mobility without getting feedback or being administered from a central controller.

© Springer Nature Switzerland AG 2021
M. H. Rehmani, *Blockchain Systems and Communication Networks: From Concepts to Implementation*, Textbooks in Telecommunication Engineering,
https://doi.org/10.1007/978-3-030-71788-9_7

Table 7.1 Different functionalities achieved in wireless networks through blockchain

Type of network	Purpose for which blockchain is used
IoT	Identity management
IoT	Authentication
IoT	Wireless power transfer
Green IoT	Secure wireless power transfer
Cellular Network	Grant free access
Cellular Network	Roaming
CRN	Channel hopping sequence
CRN	Dynamic spectrum sharing
CRN	Spectrum auction
CRN	Spectrum sensing as a service
CRN	Spectrum trading (with fine granularity)
CRN	Future spectrum trading
CRN	Spectrum patrolling
CRN	Spectrum regulation and auditing
CRN	Collision free communication
CRN	Medium access by CR nodes as an spectrum auction
Consumer Electronic Devices	Forming an authentic network (smart home)
Fog-RAN	Reaching consensus
IoV	Data sharing
IoV	Traffic management
IoV	Carpooling
SDN	Secure data transferring
Multimedia networks	Video integrity and fake video tracking
Multimedia networks	Auditing of video content
Multimedia networks	Smart contracts for video content
Multimedia networks	Peer-to-Peer video content sharing
Multimedia networks	Privacy of video content
Multimedia networks	New ways for revenue generation through video contents
Smart Grid	Energy trading
Smart Grid	Arbitrage
V2G	Energy trading
Cloud of Things	Resource management and track resource utilization
Wi-Fi	Anonymous access control
Wi-Fi	Congestion avoidance

Table 7.2 Similarities between different communication networks and DLT

Communication network	Communication network feature		DLT feature	
	Distributed	No central entity	Distributed	No central entity
MANET	Yes	Yes	Yes	Yes
D2D	Yes	Yes	Yes	Yes
CRN	Yes	Yes	Yes	Yes
Machine Type Communication	Yes	Yes	Yes	Yes
CRSN (Ad hoc mode)	Yes	Yes	Yes	Yes
UWSN (Ad hoc mode)	Yes	Yes	Yes	Yes
WSAN (Ad hoc mode)	Yes	Yes	Yes	Yes

7.1 Blockchain and Internet of Things (IoT)

In some wireless networks, the number of devices connected to the network grows quickly. For instance, Internet of Things (IoT) is one such network where an unlimited number of devices can be connected to the Internet and form a network, that is, IoT network. Devices with such a scale need proper management for their operation. These devices in IoT network are managed with the help of a centralized entity. This brings two issues in the IoT network. The first issue is scalability issue that how to manage these enormous number of devices connected to the network in an efficient manner. The second issue is the single point of failure as there is a single central entity to manage such a network.

To cope with these issues, blockchain technology can play a vital role by providing its decentralized nature of achieving consensus by incorporating a trust layer. In this manner, IoT devices can be authenticated without a centralized authority. Without a centralized entity, DLT can support such types of networks to coordinate and manage among devices located geographically at distant locations.

Figure 7.1 shows how different devices can authenticate each other without a central entity in an IoT network. More precisely, once a device is registered in an IoT network, it can communicate and transfer data with another IoT device securely. First, this newly IoT device can be authenticated by other IoT devices without the presence of a central entity. Second, when this IoT device makes a transaction with another IoT device, this operation can be confirmed by peer IoT devices through the consensus algorithm. In this manner, a transaction can be taken place between two untrusted IoT devices without going to the central entity.

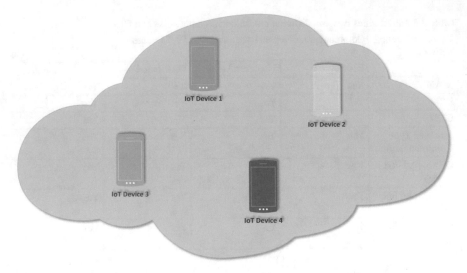

Fig. 7.1 IoT network without a central entity in which devices authenticate each other. In this figure, four IoT devices, i.e., IoT device 1-IoT device 4 are present. IoT device 1 newly joins the networks, thus it will be authenticated by other IoT device

7.2 Blockchain for Fog-RAN

In a Fog-RAN architecture, mobile devices can be used to devote some portion of their computing capabilities to assist RAN. This can be achieved by disaggregating RAN operations to these mobile devices. The number of these mobile devices can jointly execute the operations of RAN. These devices can be logically considered as a single server, where each device shares some portion of its resources, whether it can be computing, storage, or network capability, to serve the overall operation of the network. Blockchain can also help to establish such type of Fog-RAN network architecture by achieving consensus at different levels among these nodes. Moreover, blockchain can also help this type of network by using its intrinsic data storage capability to support network operations.

7.3 Blockchain and IoT Edge

When the concept of IoT was introduced, it was envisaged that cloud computing will be an integral part of the operation for IoT. More precisely, IoT devices will be connected to the cloud and all the data will be stored in cloud. Similarly, any big data processing will be carried out on the cloud as the cloud will have strong computational power, storage, and other necessary resources. However, as the IoT paradigm evolved, new applications and services were sought. Moreover, with the

Table 7.3 Cloud-based Blockchain Services

Microsoft Azure	Blockchain as a Service
Clouding Hashing	Bitcoin mining as a Service (BMaaS)
IBM Watson IoT Platform	To manage IoT data on ledger using IBM cloud service

innovation and development in machine learning paradigm, and to further reduce any delay in getting a response back from the server, researchers thought to move the resources from the cloud to the edge. This led to the IoT-enabled edge computing paradigm. This IoT-enabled edge computing has some clear advantages over IoT over the cloud paradigm. For instance, IoT devices need not wait for a long time to get their tasks done by the servers over the cloud. Instead, IoT devices can use the resources over the edge. This strategy can help in meeting delay requirements of time-sensitive applications.

Consider a scenario in which mobile users are present. These mobile users want to run different blockchain applications. Since these mobile users have limited resources, therefore, these mobile users can get benefit from the edge computing paradigm.

Imagine a scenario where there are thousands of distributed IoT devices using services and resources provided by different edge service providers. In this scenario, authenticating IoT devices and managing trust layer will be a bit challenging. Blockchain, designed to work in a decentralized environment and building trust layer in an untrusted environment, is the best fit for IoT-enabled edge computing. In IoT-enabled edge, blockchain can serve to provide a trust layer, enhanced security, and managing the records in the form of a distributed ledger.

Computing capabilities are moving from the cloud to the IoT edge. This IoT edge, however, does not provide privacy preservation of user's data. Additionally, it lacks security mechanisms as well.

When thinking about deploying blockchain for IoT devices, issues such as energy and computing constrained devices, need to be considered. Edge computing can help in this context by leveraging resources at the edge to facilitate IoT devices using blockchain. This can be in the form of helping in the mining process (or solve high computational PoW puzzle), encryption, or to perform hashing or help in authenticating IoT devices. More precisely, IoT devices ask the miners over the edger servers to solve the puzzle for it and as a result, get some payment from IoT devices. Table 7.3 shows cloud-based blockchain services. Table 7.4 shows the comparison of cloud and edge-based blockchain.

Table 7.4 Comparison of Cloud- and Edge-Based Blockchain

	Cloud-based blockchain	Edge-based blockchain
Mining	Over the cloud	Over the edge
Delay	High	Low
Proximity	Far	Close
Computer power of servers	Very high	Comparably low

7.3.1 Challenges in Blockchain-Based IoT Edge

There are few challenges as well when thinking about blockchain-based IoT edge. For instance, transaction speed of public blockchain is quite limited. In the context of IoT edge, the same issue will arise. IoT devices over the edge want to use public blockchain, thus Tx speed needs to be considered. Second issue is related to data storage. If all the IoT devices intend to store every data in the form of transactions to the blockchain, this will tremendously increase the amount of data to be stored in the ledger. This will have few effects:

1. The overall ledger storage size will increase and IoT devices may not have such storage capacity to keep the complete copy of the ledger.
2. When there are lot of IoT devices want to include their transactions, lot of blocks need to be added to the blockchain. This will increase the difficulty level of the public blockchain as IoT devices want to include more and more blocks and solving the puzzle (in case of PoW consensus) quickly.
3. Scalability will be another challenge that how massive IoT devices will be part of the ledger update process?
4. In the context of public blockchain, privacy will be the key concern that needs to be considered.
5. In a completely public blockchain, with massive IoT devices, how consensus protocol will perform working? Moreover, questions such as once the block is added to the blockchain, how quickly this information will be disseminated in IoT edge environment?
6. IoT devices are not computationally powerful so will IoT devices run the consensus or even the consensus protocol can be executed over the edge server?

7.4 Blockchain, IoT, and Consumer Electronics

Around us, we find several home appliances and devices which we use on daily basis. These devices include mobile phones (smart phones), TV, Cameras, tablets, iPods, laptops, and computers—commonly known as consumer electronics (CE) devices. These CE devices together with other IoT devices (embedded sensor in other devices) need to connect to form a network—making a smart home environment. This smart home network also needs to connect to the Internet to operate efficiently and to send and store daily usage and other statistics.

Generally, these IoT and CE devices are not too much powerful in terms of security and pose new threats and vulnerabilities when connected to the Internet. Moreover, depending upon the user, these devices may also face different levels of security attacks, as different users have different styles to operate the same device. Some users are more careful, while others are less careful, for instance, a young child may click on any link just for the sake of curiosity and thus make his device vulnerable. These devices are in massive quantity and store personal information. Securing and safeguarding these devices is important. Privacy preservation techniques (differential privacy, etc.) along with blockchain can help to secure these devices.

7.4.1 How to Manage IoT and CE Massive Data?

There are two possible ways to manage massive data generated by IoT and CE devices over the blockchain network.

7.4.1.1 Sidechains

One possible solution to manage massive data generated by the IoT and CE devices is the usage of sidechains.

7.4.1.2 Use of Multiple Blockchains

For IoT and CE devices, several blockchain networks can run in parallel depending upon the targeted applications. These different blockchains can talk to each other to share information, however, issues such as interoperability among different blockchain systems need to be considered.

7.4.2 Which Blockchain to Use for CE and IoT Devices?

There are certain advantages and disadvantages of blockchain architecture adopted for CE and IoT devices. For instance, adopting public blockchain makes the blockchain system available and accessible to everyone but managing such a huge public blockchain network may bring data management and privacy issues. On the other hand, if private blockchain system is adopted then it will be accessible only by a limited number of users but may be less secure as compared to public blockchain (from the perspective of a number of nodes to reach consensus—thus private blockchain has less nodes, there is more probability of being less tamper-proof).

7.5 Blockchain and Wireless Power Transfer—Green IoT

Wireless power transfer and energy harvesting in IoT devices are important to meet the energy requirements of these battery-powered energy-constrained IoT devices. Dedicated energy harvesters are deployed to wireless transfer the energy to these devices. In this energy transfer phase, the energy harvester first acquires the current energy state of the devices and, depending upon the pre-defined limit, energy is transferred wirelessly to these devices. However, it may possible that some malicious nodes try to deceive the harvester nodes by communicating the wrong energy state information of nodes. This may result in not delivering the correct amount of energy to the nodes that need energy and thus resulting in the non-functionality of the IoT network. It may also possible that some malicious nodes may steal some wirelessly transferred energy by declaring themselves as legitimate nodes. Another important issue is which type of energy to transfer? Means, energy generated from fossil fuels or energy generated from renewable energy resources? And how to track those energy resources? Similarly, in IoT environment, if some powerful nodes want to sell their surplus energy to nearby nodes which require energy, then how such an energy transfer is possible in a trustable environment without a centralized entity? To answer all these questions, blockchain may play a vital role. Blockchain can not only help in realizing such a wireless transfer scenario but it can also make such a network more secure and more resilient to cyber attacks and energy attacks.

7.6 Blockchain and Internet of Vehicles (IoV)

Internet of vehicles (IoV) is an emerging communication network in which vehicles are connected with the Internet and access different services to improve the overall road experience. In IoV, vehicles are connected to the roadside unit (RSUs) to access different services and to share their collected data. This results in the overall improvement in road experience and safe critical lives in road accidents. In IoT, vehicles tend to share their critical information. This information is passed to the RSU. In case of cyber-attacks, the RSUs are the single point of failure and may result in discontinuity in the network.

Blockchain is one of the viable solutions to establish trust among different vehicles in IoV. It decreases the dependency of vehicles for credential checking through RSU. In fact, blockchain, through its inherent features of decentralization and consensus capability, may establish a trust layer among the non-trusted vehicles in IoV. However, a lightweight consensus algorithm is required in terms of computation power and finality of the blocks. PoW consensus may not be a suitable solution for IoV. PoS consensus algorithm may be a good choice, as PoS algorithms do not require high computational power and the finality of blocks can be done quickly. There are several applications where blockchain can be applied in IoV. Some of them are discussed below:

- Carpooling or riding services are very much popular these days. A customer can see the available rides within a geographic location and may find a cheap ride. Similarly, the car rider or drivers may publish their location and find an appropriate nearby passenger. All these transactions carried out can be done through blockchain.
- Data sharing among the vehicles within IoV network or between RSUs and vehicles can be done in a trusted environment through blockchain.

7.7 Blockchain, Software Defined Networks (SDN), and Virtualization

Blockchain can be applied to software-defined network-based networks. There are several SDN-based networks such as SDN-based SG, SDN-based IoT, SDN-based IIoT, and SDN-based VANETs.

7.7.1 Blockchain-Based SDN: Advantages

Blockchain-based SDN has certain advantages; below, we mention few:

- Data is securely saved and transmitted.
- Hashing and encryption can be easily applied as it is the integral factor of blockchain.
- In SDN-based VANETs, blockchain-based SDN can be used to store vehicles generated data and audit any data for resolving any traffic accidents or crime. Videos can also be checked and stored, and their integrity can be confirmed.
- IoT devices can be authenticated through blockchain in SDN-based IoT.
- No need of trusted third party required in SDN-based IoT by using blockchain.

7.7.2 Virtualization, Cloud Computing, Edge, and Fog Computing

Virtualization is a process in which resources can be viewed logically instead of their actual physical appearance. These resources can be networking resources, computing resources, storage resources, or hardware resources. Through virtualization, abstraction of these resources can be achieved which will help in performance improvement and makes resource management easier.

"Cloud computing" refers to the use of Operating System (OS), networking, storage, computing, and hardware at distance. Through cloud computing, these resources are not required to be physically present at the user location, instead, these resources will be hosted by any third party, and users can access and get the benefit from these resources. In "edge computing", these resources (OS, computing, hardware, networking, and storage) will be located near to the user, thus reducing the delay incurred in accessing these resources at the cloud. In "fog computing", these resources are placed at an intermediate level, not fully at the cloud (far) and not near to the edge (near). In fog computing, these resources are placed in the middle of "cloud" and "edge". Whether it's cloud computing, edge computing, or fog computing, virtualization is the backbone of these technologies and blockchain can play a vital role in managing the resources in a virtualized environment.

7.8   Blockchain and Cloud of Things

Cloud of Things (CoT) is an emerging technology with the combination of cloud computing with IoT devices. CoT basically provides a platform to IoT devices for storing and processing their data. Traditional CoT solutions have some challenges. First, they are centralized in nature. This centralized nature of CoT results in latency issues for IoT devices. Moreover, the centralization solution also results in trusting the centralized trusted third party for storing and processing IoT big data. This raised serious privacy and security concerns. Moreover, the centralized CoT architecture also hinders the performance of the overall system as they are not scalable as IoT systems required. CoT can be hugely benefited with blockchain technology.

By incorporating blockchain with CoT, lot of benefits can be achieved such as blockchain as a service (BaaS) can be provided. Moreover, blockchain can help cloud computing companies to store the transactions and track the resources utilized by the blockchain users.

In summary, integrating blockchain with CoT will be beneficial for both technology operators as blockchain can provide its inherent features of transparency, immutability, and decentralization to CoT, thus resulting in new business models for CoT operators. Moreover, complex blockchain operations such as storing huge ledger copies can be easily done by using cloud resources and mining operations can also be performed by the cloud servers, thus reducing the load to the IoT devices, resulting in less energy consumption of IoT devices.

7.9   Blockchain in Cellular Networks

There are numerous applications of blockchain for mobile communication networks. Some of these applications are mentioned below:

- Spectrum sharing/trading/leasing
- Telecom infrastructure usage by other companies
- Network slicing
- Cloud-related applications
- Data offloading
- Access control
- IoT devices Identity Management
- Detection of malicious software and applications
- Mobile crowdsourcing

7.9.1   Blockchain and Mobile Devices

In traditional networks, resourceful devices are used in blockchain networks. These resourceful devices need to be memory rich and computationally powerful for storing the copy of the ledger and also to solve the computationally intensive mining puzzle. However, the role of newly innovative applications due to rich penetration of mobile devices into our daily life seems difficult due to the resource constraint nature of mobile devices (i.e., smartphones). With the expectation of newly applications by 5G and beyond 5G communication networks, these mobile devices are getting more powerful, but still, these devices cannot meet the traditional devices used over the blockchain network. Thus, in order to enable full functionalities of blockchain network using these mobile devices, an extra effort is required to re-design the basic architecture of blockchain network. For instance, the mining process needs to be lightweight; otherwise, the computationally less powerful mobile devices may not be able to solve the mining puzzle. Similarly, in terms of memory, the whole copy of the ledger cannot be stored on mobile device. Thus, alternative ways need to be investigated to integrate mobile devices in a fully functional blockchain network.

Additionally, the energy consumption of mobile devices due to the highly intensive nature of communication in blockchain network is also a major issue because mobile devices have battery issues. Therefore, the design of blockchain network should consider all these aforementioned constraints.

Recent development in the consensus protocols for blockchain network resulted in PoS-based consensus protocols. This type of consensus protocol does not require energy and computational power in the mining process such as PoW requires. Thus, the extensive computational requirement of solving PoW puzzle is not a requirement in PoS consensus protocols, making it a natural fit for mobile devices.

7.9.2 Blockchain and Roaming in Cellular Networks

In cellular networks, data roaming is one scenario in which blockchain technology can play an auspicious role. A typical roaming scenario is the one in which a mobile user goes outside the coverage region of service provider. The service provider that facilitates mobile user in a home country known as Home Network Mobile Operation (HNMO) and the service provider that facilitates mobile user in the visiting country is known as Visiting Network Mobile Operation (VNMO). Once roaming occurred; several parties are involved during the roaming process. Blockchain through its smart contract capability can help reduce not only the complex business agreements between different mobile network operators but it can improve the cellular network performance.

Roaming can be international roaming and national roaming. In international roaming, the mobile user visits any other country and uses the services provided by the visiting country. In order to have seamless connectivity and roaming, both the visiting country and home country mobile network operators need to have agreements in place to facilitate such interaction. Similarly, national roaming represents the mobility of mobile users to any other region or city where mobile user's mobile network coverage is not available. In this case, different network operators need to have an agreement with each other to facilitate such domestic roaming.

7.10 Blockchain and Wi-Fi Networks

In Wi-Fi networks, blockchain can be used to provide access control. Thus, by using the blockchain concept in Wi-Fi networks, Remote Authentication Dial-In User Service (RADIUS) server—which is used for centralized authentication—is no longer required. Similarly, the pseudo-anonymity feature of public blockchain can help Wi-Fi network service providers to protect the privacy of Wi-Fi users.

> Proof-of-Networking (PoN): This is a variant of the proof of work consensus protocol. In PoN, the messages sent by a node are considered as the "work" and then it is used to generate a new block in the blockchain.

7.11 Multimedia Communication Networks and Blockchain

Different types of multimedia content such as audio, video, images, and files are generated and communicated over the communication networks. The integrity of this multimedia content needs to be verified. For instance, in a vehicular ad hoc network, vehicles are responsible to record video streams of an accident and then provide it to law enforcement agencies for legal proceedings. These videos can play an important role in the identification of the criminal activity or sequence of events that occurred. It may be possible that these videos get forged by adding, deleting, or modifying frames in the video. This will have an adverse effect on the criminal proceedings. Thus, it is very important to test the integrity of the video which are used in such cases.

Government manages lot of records and these records can be archived and used in future. These archives can be in the form of audio recordings, video recordings, and a collection of images. The contents of these archives may contain tribunal recording, court proceedings, and proceedings of the national assembly. Blockchain can help in these archives (storages) by ensuring the integrity of the contents.

Traditionally, video integrity can be carried out using various methods. These methods can be classified as active and passive methods. In active methods, prior information such as signatures or hash values are required as proof. These methods include digital signatures and digital watermarking schemes. In passive methods, no proof of prior information is required. These methods rely on the analysis of noise in the forged video or compression artifacts. Blockchain as a potential technology can help in verifying the integrity of such multimedia content, especially video integrity verification. There can be different approaches in verifying the video integrity. One simplest approach is to apply blockchain on every frame of the video i.e., to record every frame on the blockchain. However, this seems too complex and time-consuming and may incur large processing and memory overhead. Another approach is to keep different video segments over the blockchain network, and this may reduce the processing and memory overhead.

The integrity of video stored on a blockchain can be verified by checking the hash values of the video in question against the hash values of video stored over the blockchain.

7.11.1   Video Streaming Communication Networks and Blockchain

Through video streaming service, users watch the videos as per their choice. Different services including Netflix and Amazon Prime Videos are available to facilitate video on-demand. In addition to this, video streaming services based on blockchain are also available such as Livepeer.

> Multimedia communication is essential in contemporary networks.

Blockchain can be very helpful in video applications. The content of video can now be generated by an ordinary person and can be posted on YouTube, TikTok, Facebook, and Twitter. Video blogs or video logs (vlogs) and video contents are created for travel, cooking, marriage events, and even people produce and share their living habits through these videos.

7.11.2 New Methods of Revenue Generation and Business Models

New business models and revenue generation methods are developed. Previously, only the video content generator was rewarded according to the number of likes, number of times the video is watched, and advertising links attached to the video content but now even the video content viewers can get incentive and receive rewards and this is available through platforms like Bonzo and Me. Blockchain can be very useful to support new business models and revenue generation for video content. Through blockchain, one can automate the business process for video contents. Moreover, payments can be done quickly, without relying the central entity, through blockchain-enabled video content.

7.11.3 Auditing for Video Content Generated Revenue

Through blockchain-enabled video content, audit is much easier. The companies can easily track the ownership of the video content and the statistics related to that video. This makes the video content system much more transparent. Similarly, the revenue can be given to the correct owner of the video content generator.

7.11.4 Smart Contracts for Video Content

Blockchain-enabled video systems will also reduce delays in payment to different parties and make the system automated through smart contracts.

7.11.5 Peer-to-Peer Video Content Sharing

Blockchain-enabled P2P video content sharing will also be possible. Now people can generate the video content and trade it without relying on third parties and this may also increase their revenue by paying less fees to these third parties.

7.11.6 Resolving of Privacy Issues Through Blockchain

Blockchain-enabled video content systems can help in resolving privacy issues. Every video generated can be recorded in a blockchain network, thus bogus claims of ownership of video content can be verified.

7.11.7 Fake Video Generation and Tracking

Through blockchain-enabled video systems, fake videos can be identified. Similarly, video disseminated on social media used for any violent purpose or harassing someone's feelings can be tracked easily. Moreover, illegal copies of the video content can be tracked and identified using blockchain and watermarking techniques.

7.11.8 Privacy of Video Content

Privacy of video content can be guaranteed by integrating privacy preservation techniques together with blockchain.

7.12 Smart Grid Communication System and Blockchain

Smart grid is an advanced form of the traditional electrical grid. Blockchain can be applied to smart grid as well and blockchain brings several advantages to smart grid communication system.

7.12.1 Prosumers

Prosumers are smart grid consumers which not only consume the energy but also produce the energy and inject back the energy into the grid. The involvement of prosumers at a large scale requires an energy market where energy trading becomes possible, thus requiring control strategies as well. These control strategies are required to balance the demand and supply gap as well as to stabilize the overall energy in the grid because the energy to be generated by the prosumer is RER-based which is intermittent in nature and not always available. Such type of energy trading system is often termed as Transactive Energy System (TES).

In blockchain-enabled transactive energy systems, one issue is related to immutable nature of blockchain. Though immutability is desired in blockchain energy systems, this immutability may also be considered as a negative aspect. The reason for this negative aspect is coding errors (in smart contracts) which may make the blockchain energy network unstable or lead to penetrate the malicious users within the blockchain system.

7.12.2 Energy Trading Benefits

Smart grid is all about electrical energy and this energy can also be traded. Through energy trading, several benefits can be achieved. For instance, renewable energy generation can be more promoted, and, in this manner, fossil fuel-based energy generation can be discouraged. Secondly, net metering (injecting back the energy into the grid) can help energy traders to generate some revenue. In order to manage such energy trading solutions, virtual power plants (VPPs) are present. VPPs basically facilitate centralized energy trading by using the traditional database. Simple database for energy trading and management in VPP is not a feasible solution as the number of smart homes connected to the SG grows, the amount of data generated is enormous, thus it will be difficult to manage. In addition, managing such a huge amount of bidding and transaction processing will be difficult and not appropriate in a centralized trading system. However, these VPPs can be further improved by considering the concept of blockchain within it.

In smart grid context and with the penetration and integration of RERs into the SG, it is still not feasible to completely rely on RERs for the operation of Micro Grids (MGs). The MGs also need to get connected with the main grid in order to get a substantial amount of fossil fuel-based energy. However, it is envisioned that after a decade or so, one may completely rely on RERs. In this manner, prosumers in MGs or Networked Micro Grids (NMGs) can perform energy trading locally and in an autonomous manner. DLT or more specifically blockchain technology can play an important role by keeping track of energy consumed and produced in a transparent manner. Moreover, blockchain can also keep track of energy prices and make the bidding process efficient.

"SolarCoin" is a blockchain-based energy trading system available online. In this blockchain system, a prosumer who generates the energy can first register itself to the SolarCoin and after validating its solar generation capacity and equipment, a SolarCoin is issued to the prosumer based upon the generated renewable energy. SolarCoin can be accessed from here: https://solarcoin.org/.

7.12.3 Privacy Preservation in Blockchain-Enabled Smart Grid

There are various sources of big data generation these days ranging from social networks to smart homes. This huge amount of big data, if exposed, make the user's privacy more vulnerable. In order to preserve privacy, there are several privacy preservation techniques such as generalization, suppression, swapping, buketization, and randominzation. To summarize, the goal of privacy preservation algorithms is to obfuscate smart grid-related parameters to the adversaries. Differential privacy is an example of a privacy preservation algorithm that comes under the category of randomization.

7.12.3.1 Diffential Privacy

Differential privacy is proposed by Cynthia Dwork and it has been applied widely in communication networks along with blockchain to obfuscate information about an individual person or identity and this is achieved by adding noise in a controlled manner to the data. By applying differential privacy, we can also control the level of privacy, referred hereafter as "privacy budget", and it is denoted and controlled by the parameter epsilon ϵ. If we want to achieve strong privacy, we should have low values of epsilon and vice versa. When applying differential privacy to any scenario, we also need to consider how much obfuscation is carried out so that we also preserve the realism of the considered scenario. In simple words, noise should be added in such a way that it obfuscates the data and the true meaning of data should not be lost and it can be used in a purposeful manner.

In differential privacy, random noise is added using Laplacian or Exponential mechanism. In DP, a tradeoff exists, i.e., the tradeoff between utility and privacy. By adding noise, more and more privacy can be achieved but the utility will decrease. Utility means the availability or realism of actual data. More precisely, utility describes us that after applying privacy preservation algorithm, how accurate and correct the data is?

This addition of noise to the original data can be done in two ways. It can be done in real time, i.e., online. More specifically, when someone makes a query to the original data, then noise is added to the data and then this noisy data is provided. This is known as online or interactive mode. It can also be done as an offline mode, i.e., as soon as this data is available, noise is added to it, irrespective of when someone makes a query. This is known as a non-interactive (offline) mode.

> The basic aim of differential privacy is to publish information to the requester without disclosing the real record of an individual person or entity.

7.12.3.2   Smart Grid Environment and Privacy

In a smart grid environment, customers can be divided into two major categories. The first one is domestic users and the second one is industrial users. For both types of users, preserving privacy is important. Privacy preservation algorithms can be applied in various ways to guarantee the privacy of users. For industrial users, privacy also matters as it will disclose the operating capacity of industrial customers. Similarly, in some industrial units, they have their own power generating units which are also connected with the main grid. Exposing these parameters to an adversary may also cause harm to the industrial users. In a broader context, the utility itself wants to preserve the privacy of its operational parameters such as the exact location of grid and power stations, location of transformers, different parameters associated with their grid including frequency, power factor, etc. Revealing all such information to the adversary may result in exposing how the utility network operates. On the other hand, industrial users and utilities also want to share this information with third parties for analysis. Moreover, such type of data also needs to share within different parties involved within the smart grid for different purposes such as demand response and real time price adjustment. Thus, sharing this information is necessary while also preserving the privacy of the customers.

> "Proof of Energy" is a consensus protocol which is based on PoS and considers consumption and production of energy by the prosumer which selects the miner node.

7.12.3.3   Smart Homes and Privacy Preservation

Smart homes will be an integral part of smart grid. In smart homes, all the home appliances will be connected through communication technologies. These smart home appliances can be monitored and controlled through smart meter. Smart meter is the advanced version of traditional electric meters through which one can monitor the overall electricity usage of a smart home. This monitoring is not just limited to the overall consumption of electricity, instead, a more fine-grained monitoring at appliance level in real time is possible.

There are multiple reasons for which this data is collected. For instance, through this data, one can see real time electricity usage of appliances and thus can identify any appliance which is consuming more energy and can be stopped working at any particular time or replaced with a more energy-efficient appliance. Similarly, billing can be optimized by scheduling the operating time of appliances, thus appliances consuming higher energy can be shifted to less peak hours (less tariff) and thus revenue savings.

With this fine-grained monitoring, information about an individual's habits can be revealed. Using this monitored data, an adversary may also understand the usage level of each appliance, its activities during the day and night period and also when a particular task is happening within a smart home, for instance by looking at toaster, one may conclude that someone is using the kitchen in the smart home and taking breakfast. In this manner, customer's behavior can be revealed, and this information may be used by third parties, let's say, for advertising campaigns.

Similarly, with smart meters and communication technologies, remote monitoring of home appliances and their switch on/off decision can be taken place. Additionally, any load scheduling can also be notified with the help of smart meters. Smart meters can also play their role in net metering (i.e., injecting back the generated power to the grid).

7.12.4 Vehicle to Grid (V2G) Energy Trading

Electric vehicles can be considered as moving power plants or moving storage devices. Imagine a situation in which there is an excessive amount of energy being generated by the generators in the grid. This excessive energy can be in the form of renewable energy or fossil fuel-based energy and will be stored by hundreds of electric vehicles. This fleet of electric vehicles moves from one place to another, from one city to another. If there is any shortage of energy, then the fleet of this EV can play its role and inject back the energy into the grid. This complete paradigm is known as Vehicle to Grid (V2G) network. Energy trading in V2G network can be tracked and performed easily with the help of blockchain. It is also possible that since each vehicle may not have computational power, it may offload the mining process to the edge computing server (may be a roadside unit (RSU)). This offloading process may ease off the burden from EVs to the RSUs. To summarize, EVs can have two major roles in smart grid from the perspective of energy trading. EVs can inject back the energy into the grid to balance the supply and demand gap of the smart grid – acting as energy producers. Additionally, EVs can store the excessive energy during less peak hours and serve as consumers and thus utilize it for its own purpose.

7.12.5   Effect of DoS on Energy Trading Market

In a blockchain-enabled energy trading system in smart grid, attacks may have an adverse effect on the overall blockchain system. For instance, a DDoS attack can severely impact the demand supply gap by stopping the bids from reacting to the market, thus predicting a wrong load curve. This may also have an effect on the available energy market price. Moreover, since the bids will not reach the market and energy selling transactions will not occur, then this will result in de-motivating the prosumers to participate in the energy trading market. This will ultimately lead to instability in the energy market.

7.12.6   Cryptocurrency in Energy Trading Systems

Cryptocurrency such as Bitcoin and Ethereum are very famous and with the passage of time, new cryptocurrencies are getting introduced. These cryptocurrencies can be used and exchange as financial transactions and can be served as tokens in different markets. One such market is energy trading market where cryptocurrency can be used in exchange for trading of electricity.

The use of these cryptocurrencies in energy trading markets have some advantages such as there will not be any need of a centralized third party, however, the processing speed of the current cryptocurrency-based systems (e.g., Bitcoin) is much slower to compete with online credit card transaction systems. The advantages of these cryptocurrencies are much higher than their disadvantages, thus, these cryptocurrencies can be used substantially in energy trading system.

Cryptocurrency such as Bitcoin is basically a public blockchain system which consumes lot of energy and this energy is consumed due to mining process. Moreover, due to the public nature of Bitcoin and dependency on PoW consensus algorithm, the transaction speed is slower. If both of these challenges are handled carefully, it is evident that cryptocurrency can play an important role in future energy market trading system.

7.12.7   Arbitrage in Energy Trading Systems/Markets
Through Blockchain Systems

Arbitrage is a process in which an entity buys and sells something simultaneously. The buying will be carried out at lower price and selling is done at higher price, thus resulting in net profit to the entity without net investment. Arbitrage in energy trading system/markets is also possible and can be carried out among different commodities within the energy market. This energy arbitrage can also be supported through blockchain-based systems.[1]

7.12.8 Renewable Energy Resources and Negative Pricing

RERs are highly intermittent in nature and unreliable. Due to high penetration in the SG and the plan to increase this penetration to a level to an extent where energy generated through RERs surpass the conventional power generation methods, will have a dual impact on the grid; its instability and negative pricing. In fact, when both the traditional power generation and RER-based generation occur at the same time then there is a chance that the demand and supply of electricity within the SG impacts. More precisely, when the demand increases and the supply decreases, this will result in an increase in prices. On the contrary, when the demand decrease, and the supply increase, this will result in a decrease in price. And it may lead to a condition where the traditional generation cannot stop their energy generation due to economic reasons and RERs prosumers continuously add the generated energy to the SG. This huge demand supply gap in the energy market will result in a negative pricing phenomenon in the energy trading market, i.e., power generating units sell the energy units without any benefit or even selling their energy units at garbage value. This phenomenon of negative energy pricing is similar to the international oil market where oil companies just sell their oil at negative prices.

As reported in the literature, there are few ways to handle this issue of negative pricing in the energy trading market. One obvious way is to incentivize the energy users to use DR programs so that the DR energy curve can be adjusted to avoid any financial as well as energy losses. Another direction pointed out in the literature is to use the blockchain's mining process as a method to balance the demand response energy curve, i.e., to ask the mining nodes to perform their mining when energy is available in abundance. This will have a dual impact on the DR energy curve i.e., energy can be balanced and the energy price per unit will also decrease during the mining process. The energy consumption during the mining process of blockchain-based energy trading system can be further reduced if the architecture is modified, i.e., instead of using public blockchain may be consortium blockchain can be used. This will reduce energy consumption during PoW consensus and Tx speed can be enhanced.

When using blockchain and transfer energy through P2P energy trading, the amount of energy is small. For each transfer of energy from one physical location to another, losses can occur due to inverter involvement. Thus, this aspect needs to be considered. Similarly, for a huge amount of energy trading, lot of communication

---

[1] Islamic commandments are present on arbitrage and those who are interested to further investigate and find the rulings on this, the preliminary book on Islamic Finance by Hazrat Mufti Muhammad Taqi Usmani Sahib Damat Barakatuhum can be found here [88].

is required, thus efficient communication protocols are required which incur less overhead in maintaining these transactions as a ledger.

7.13 Communication Networks and the Use of Blockchain with Machine Learning

7.13.1 Machine Learning and Communication Networks

Machine learning has been used in different application domains ranging from medical image analysis to speech recognition. In communication networks, machine learning has been widely adopted to reduce human participation to analyze the massive amount of data, classify different types of network traffic, and monitor the communication system to identify anomalies or intrusion detection. Machine learning-based communication systems have the capacity to handle big data applications and is a strong candidate for 5G and 6G communications networks to optimize the system performance.

7.13.2 Classification of Machine Learning Techniques and Blockchain

Machine learning (ML) techniques can be classified as unsupervised techniques, supervised techniques, and reinforcement learning techniques. Each of these categories of ML techniques has its own requirements. For instance, supervised ML techniques need a huge amount of data set for training purpose. In unsupervised ML techniques, no training data set is required. Reinforcement learning techniques interact with the environment and learn and react accordingly. As mentioned before, due to the vast applications and benefits of ML, it has been applied to communication networks. Now, with the integration of blockchain in such scenarios, what benefits can be achieved? And what challenges one can face? These are the questions that are under investigation by researchers around the globe. When ML-based techniques applied to blockchain-enabled communication networks, lot of benefits can be achieved. For instance, the data sets or data being used as an input for ML algorithm can be more trustable and secure. DApps can be more intelligent by using ML techniques and smart contracts can be further enhanced and automated by incorporating ML algorithms.

7.13.2.1   Supevised ML

In supervised ML, labeled training data set is available, and the algorithm learns from it. Famous supervised ML learning algorithms include support vector machine (SVM), k-nearest neighbor (kNN), decision tree, and neural networks. Supervised ML algorithms can be used in blockchain-enabled communication networks for intrusion detection and network traffic classification used for various purposes.

7.13.2.2   Unsupervised ML

In unsupervised ML, labeled data set is not available and ML algorithm is required to extract features from the data. Such type of ML algorithms can be used in blockchain-enabled communication networks for the identification of any type of clusters, for instance, of nodes generating a specific sort of traffic or launching a DoS attack. Famous unsupervised ML algorithms are principal component analysis (PCA) and k-means clustering algorithm.

7.13.2.3   Reinforcement Learning Algorithms

In the reinforcement learning (RL) algorithm, feedback from the environment is required for the algorithm to learn and adapt. Such type of RL algorithms have been applied to radio resource management, link failure detection, and reconfiguration of networks. Famous algorithms in this category are multi-armed bandit (MAB), Q-learning, and Markov decision process.

7.13.3   Advantages of Using Machine Learning in Blockchain-Enabled Communications Networks

Blockchain inherent features can help a lot in deploying ML algorithms in communication systems. For instance, the immutability feature of blockchain reduced the chances of data tampering, thus ML algorithms can rely on such data for training and testing purposes. Moreover, since blockchain provides auditability, therefore, ML algorithms when training their algorithms can also get benefit of this auditability feature and one can easily backtrack the training data used for ML algorithms. In the context of IoT devices, blockchain provides identity management and allows the devices to securely become part of the blockchain network. Thus, relying on ML algorithms can also trust the IoT device and rely on their data generated and use different distributed IoT devices for decentralized training of ML algorithm. However, certainly, there are few challenges as well for ML in these systems. Metadata is not secure in blockchain privacy leakage. Moreover, with more advanced communication applications and incorporation of massive communication devices, ML algorithms will find difficulty in terms of scalability as data sets will increase enormously. Additionally, memory and computation constrained IoT devices will be finding it difficult to execute ML algorithms and get benefit from this massive data for training purpose. Federated learning is one such solution in such a scenario. Blockchain-based communication networks also used federated learning techniques to handle such issues.

ML algorithms can also enhance the core functionalities of blockchain-enabled communication systems. For instance, consensus algorithms can be further improved by using ML algorithms. More precisely, leader election (mining process) can be carried out quickly and efficiently. Additionally, any misbehaving nodes can be identified by incorporating ML algorithms during the mining process. Another important aspect ML algorithm can bring to blockchain-enabled communication networks is the identification of bugs in the smart contracts. Thus, by using ML algorithms in smart contracts, secure and safer smart contracts can be designed and implemented in blockchain-enabled communication networks.

7.14 Summary

This is the last chapter of this book. In this chapter, we discussed in detail the applications of blockchain systems in different communication networks. We first presented how blockchain can be applied to the Internet of Things and its related technologies, such as edge computing, cloud computing, Fog-RAN, and green IoT. We then discussed how blockchain can be helpful in addressing the needs of consumer electronics. Internet of vehicles and the use of blockchain technology were then discussed. Software-defined networks and virtualization were then discussed from the perspective of blockchain. We discussed in detail the applications of blockchain in cellular networks including roaming and the role of mobile devices within blockchain. Multimedia communication and blockchain were also discussed in detail. Smart grid communication networks and blockchain systems are also discussed in detail. And finally, we ended the chapter with discussion of applications of machine learning to blockchain systems.

7.15 Future Research Directions

In this section, we highlight few future research directions related to communication networks and blockchain.

- How central entities in IoT network scale as the number of devices grow?
- How to manage a huge number of devices in a centralized IoT network?
- The centralized IoT network can be considered as a single point of failure?

- For delay-sensitive wireless communication network applications, how PoW will be used?
- To investigate optimizing PoW or propose new consensus protocols which have low validating time for approving transactions.
- How much energy can be saved if wireless energy harvesters use renewable energy instead of traditional fossil fuel-based energy?
- In case of energy-related attacks by malicious users and countering them effectively, how much energy can be saved?

7.15.1 Blockchain, Smart Grid, and Peer-to-Peer Energy Trading

In a blockchain-based peer-to-peer energy trading system, the availability of real-time information about the underlying grid, the quality of power being transferred, the amount of loss that may happen due to conditions of power lines, and overall utility network power imbalance can be considered when performing P2P trading.

Another interesting aspect is how much energy a peer can transfer or inject or traded in the network? Will it imbalance the network? Is there any upper limit on energy traded by the peers?

In the P2P energy trading system, it will be difficult to regulate the energy trading. The reason for this is that if this regulation will happen then which entity will do that and how this entity ensures that the energy being traded is physically transferred? And how much energy was lost during the physical delivery of the energy among the peer nodes? Moreover, it is also required to capture what benefit will the utility itself get when helps to transfer the traded energy physically? Will it charge the peer nodes for this transfer? Will the utility charge from the sending peer or both?

7.16 Further Reading

This section outlines further readings.

7.16.1 Blockchain and IoT, Edge, Fog, and Cloud Computing

General discussion on IoT can be found in [50]. There has been lot of work done in the domain of blockchain for IoT, edge, fog, and cloud computing. Recently a book is published on edge computing and blockchain [73]. A pioneering article on applications of blockchain and IoT is [9]. Some wonderful resources on blockchain and IoT are [8, 99]. Blockchain, virtualization, and SDN are discussed in [5, 102].

Standardization efforts on blockchain and cloud computing can be found in [87]. A nice discussion on blockchain and cloud of things can be found in [67]. Wireless power transfer [3] and blockchain with the integration of IoT can be found in [44]. A general discussion on blockchain security is available in [52].

7.16.2 Blockchain, Wi-Fi, and Mobile Communication

General discussion about wireless networks can be found in [2, 23, 77]. A discussion on Wi-Fi and mobile communication using blockchain can be found in [21, 57, 68, 72].

7.16.3 Smart Grid and Blockchain

A very nice book on smart grid is [63]. Some other relevant literature are [74, 75]. General discussion on differential privacy, IoT, smart grid, and RERs can be found in [13, 36, 39–41, 78, 79]. Big data and privacy preservation from the perspective of communication is discussed in [91]. Lot of work has been done on blockchain and differential privacy [37, 38].

Privacy preservation and differential privacy using blockchain for energy trading in smart grid can be found in [14, 31, 32, 101, 105]. There is excellent work available on blockchain, energy trading, and smart grid [86, 100]. Peer-to-peer transactive energy in blockchain is discussed in [83]. Other wonderful resources on blockchain-enabled energy trading are [27, 34, 85, 104].

7.16.4 Multimedia and Blockchain

Effective capacity in wireless networks is discussed in [11]. Full duplex and multimedia communication for cognitive radio networks is discussed in [10, 12]. Medical image analysis (in which blockchain can also be applied for checking the integrity of medical images) can be found in [6, 7]. Other wonderful resources on blockchain and video streaming, and tampering can be found in [15, 22, 33].

7.16.5 Blockchain, Machine Learning, and Communication Networks

Blockchain and Machine learning for Communications and Networking Systems is discussed in detail in [60].

Problems

7.1 Explain how blockchain systems can be used in communication networks.

7.2 What cloud based blockchain services are available?

7.3 Describe few challenges in blockchain-enabled IoT-edge.

7.4 How blockchain can be beneficial for consumer electronics?

7.5 How blockchain can be used to manage big data?

7.6 Mention few applications of blockchain in cellular networks.

7.7 Define prosumers.

7.8 Different cloud based blockchain services are available by different companies. Explore and compare how much these companies charge for providing and hosting blockchain solutions with different configurations of nodes, memory, and computation resources.

7.9 Investigate how much memory is available in different IoT devices and how practically these IoT devices can support blockchain code (smart contracts, copy of ledgers, and mining capability).

7.10 Consider a blockchain enabled roaming scenario in cellular network. A user travel from home country to visiting country. The home country and the visiting country, both have the cost and the revenue. Consider flat-rate prices, investigate how much gain can be achieved in terms of revenue by using this blockchain system compared with traditional roaming scenario.

7.11 Consider a blockchain enabled energy trading system where privacy preservation algorithms have been used to secure the privacy of users. Investigate how these privacy preservation algorithms can preserve the privacy of traders while maintaining the workability of the trading system.

7.12 Investigate the impact of negative pricing in a blockchain enabled energy trading system considering different types of RERs such as solar and wind.

Solutions

In this part, we provide solutions of challenging questions of the respective chapters.

Problems of Chap. 2

2.1 Let's take $N = 20$. This blockchain network can tolerate 3 to 9 faulty nodes.

$$[\frac{N-1}{5}] = [\frac{20-1}{5}] = \frac{19}{5} = 3.8 = 3, \tag{7.1}$$

$$[\frac{N-1}{2}] = [\frac{20-1}{2}] = \frac{19}{2} = 9.5 = 9, \tag{7.2}$$

2.2 Let's explore how to find the size of Ethereum blockchain. Visit these websites for finding the current size of Ethereum blockchain:
https://etherscan.io/chartsync/chaindefault
https://blockchair.com/ethereum/charts/blockchain-size

2.3 Let's explore how to find the transaction handling capacity of blockchain. Transaction rate per second can be found on this website:
https://www.blockchain.com/charts/transactions-per-second.

Problems of Chap. 3

3.1 As required in the problem, we need to use the basic commands of Ubuntu OS, thus, follow the following procedure for key generation and using sha384 and md5 hashing algorithms.

- We use "ls" command to display the list of files and directories.
- We use "rm" command to delete the file.
- We use "echo" command to write the passed arguments to the standard output.
- We use "cat" command to concatenate files, create or view files, and to redirect output to the files.

© Springer Nature Switzerland AG 2021
M. H. Rehmani, *Blockchain Systems and Communication Networks: From Concepts to Implementation*, Textbooks in Telecommunication Engineering,
https://doi.org/10.1007/978-3-030-71788-9

- We use "gpg" to generate an OpenPGP key. More details can be find in the link below: https://help.ubuntu.com/community/GnuPrivacyGuardHowto
- We use "shasum" to print and check SHA checksums. More details can be find in the link below: https://manpages.ubuntu.com/manpages/trusty/man1/shasum.1.html
- We use "sha384sum" to print and check SHA384 checksums. More details can be find in the link below: https://manpages.ubuntu.com/manpages/trusty/man1/sha384sum.1.html
- We use "md5sum" to print and check md5sum checksums. More details can be find in the link below: https://manpages.ubuntu.com/manpages/trusty/man1/md5sum.1.html

Problems of Chap. 4

4.1 In the problem, we have given a blockchain network which has a block size of 4 MBytes, transaction size of 100 Bytes, Block time is 400 seconds and it requires to wait for 20 blocks to confirm any new block. We need to calculate average block confirmation time and transaction per second (throughput) of this blockchain network.

$$Transaction\ Throughput = \frac{4\ (M\ Bytes)}{100\ (Bytes) \times 400\ (seconds)}, \tag{7.3}$$

$$Transaction\ Throughput = \frac{4 \times 10^6}{100 \times 400}, \tag{7.4}$$

$$Transaction\ Throughput = 100\frac{Tx}{seconds}, \tag{7.5}$$

$$Average\ Confirmation\ Time = 20 \times 400\ seconds = 8000\ seconds, \tag{7.6}$$

The Transaction Throughput is 100 Tx/s and Average Confirmation Time is 8000 seconds.

4.2 In the problem, we have given a blockchain network with seven participating nodes $(N_1, N_2, \ldots, N_7)$ having computing power (hash rate) values $\varphi_1, \varphi_2, \varphi_3, \ldots, \varphi_7$. The computation power (hash rates) are: $\varphi_1 = 18, \varphi_2 = 14, \varphi_3 = 15, \varphi_4 = 100, \varphi_5 = 30, \varphi_6 = 16, \varphi_7 = 150$. We need to calculate the probability of wining the puzzle by each participating node.

The winning probability P_w of each participating node can be calculated as:

$$P_{w_i} = \frac{\varphi_i}{\sum\limits_{j=1}^{N} \varphi_j}, \tag{7.7}$$

where i is the participating blockchain node, N is the total number of nodes, and φ is the computation power of ith node.

The wining probability of participating node N_1 can be calculated as:

$$P_{w_1} = \frac{\varphi_1}{\sum\limits_{j=1}^{7} \varphi_j}, \tag{7.8}$$

$$P_{w_1} = \frac{18}{18 + 14 + 15 + 100 + 30 + 16 + 150}, \tag{7.9}$$

$$P_{w_1} = 0.052, \tag{7.10}$$

Similarly, the wining probability of participating node N_7 can be calculated as:

$$P_{w_7} = \frac{\varphi_7}{\sum\limits_{j=1}^{7} \varphi_j}, \tag{7.11}$$

$$P_{w_7} = \frac{150}{18 + 14 + 15 + 100 + 30 + 16 + 150}, \tag{7.12}$$

$$P_{w_7} = 0.43, \tag{7.13}$$

References

1. Ahamed, N.N., Karthikeyan, P., Anandaraj, S.P., Vignesh, R.: Sea food supply chain management using blockchain. In: 2020 6th International Conference on Advanced Computing and Communication Systems (ICACCS), pp. 473–476 (2020)
2. Ahmad, A., Ahmad, S., Rehmani, M.H., Hassan, N.U.: A survey on radio resource allocation in cognitive radio sensor networks. IEEE Commun. Surv. Tutor. **17**(2), 888–917 (2015)
3. Akhtar, F., Rehmani, M.H.: Energy harvesting for self-sustainable wireless body area networks. IT Prof. **19**(2), 32–40 (2017)
4. Akhtar, F., Rehmani, M.H., Reisslein, M.: White space: definitional perspectives and their role in exploiting spectrum opportunities. Telecommun. Policy **40**(4), 319–331 (2016)
5. Alharbi, T.: Deployment of blockchain technology in software defined networks: a survey. IEEE Access **8**, 9146–9156 (2020)
6. Ali, H., Sharif, M., Yasmin, M., Rehmani, M.H.: Color-based template selection for detection of gastric abnormalities in video endoscopy. Biomed. Signal Process. Control **56**, 101668 (2020)
7. Ali, H., Yasmin, M., Sharif, M., Rehmani, M.H., Riaz, F.: A survey of feature extraction and fusion of deep learning for detection of abnormalities in video endoscopy of gastrointestinal-tract. Springer Artif. Intell. Rev. **53**, 2635–2707 (2020)
8. Ali, M.S., Vecchio, M., Antonelli, F.: Enabling a blockchain-based IoT edge. IEEE Internet Things Mag. **1**(2), 24–29 (2018)
9. Ali, M.S., Vecchio, M., Pincheira, M., Dolui, K., Antonelli, F., Rehmani, M.H.: Applications of blockchains in the internet of things: a comprehensive survey. IEEE Commun. Surv. Tutor. **21**(2), 1676–1717 (2019)
10. Amjad, M., Akhtar, F., Rehmani, M.H., Reisslein, M., Umer, T.: Full-duplex communication in cognitive radio networks: a survey. IEEE Commun. Surv. Tutor. **19**(4), 2158–2191 (2017)
11. Amjad, M., Musavian, L., Rehmani, M.H.: Effective capacity in wireless networks: a comprehensive survey. IEEE Commun. Surv. Tutor. **21**(4), 3007–3038 (2019)
12. Amjad, M., Rehmani, M.H., Mao, S.: Wireless multimedia cognitive radio networks: a comprehensive survey. IEEE Commun. Surv. Tutor. **20**(2), 1056–1103 (2018)
13. Arshad, S., Azam, M.A., Rehmani, M.H., Loo, J.: Recent advances in information-centric networking-based internet of things (ICN-IoT). IEEE Internet Things J. **6**(2), 2128–2158 (2019)
14. Barbosa, P., Brito, A., Almeida, H.: A technique to provide differential privacy for appliance usage in smart metering. Inf. Sci. **370–371**, 355–367 (2016)

© Springer Nature Switzerland AG 2021
M. H. Rehmani, *Blockchain Systems and Communication Networks: From Concepts to Implementation*, Textbooks in Telecommunication Engineering,
https://doi.org/10.1007/978-3-030-71788-9

15. Barman, N., Deepak, G.C., Martini, M.G.: Blockchain for video streaming: opportunities, challenges, and open issues. Computer **53**(7), 45–56 (2020)
16. Basnayake, B.M.A.L., Rajapakse, C.: A blockchain-based decentralized system to ensure the transparency of organic food supply chain. In: 2019 International Research Conference on Smart Computing and Systems Engineering (SCSE), pp. 103–107 (2019)
17. Bayhan, S., Zubow, A., Gawłowicz, P., Wolisz, A.: Smart contracts for spectrum sensing as a service. IEEE Trans. Cogn. Commun. Netw. **5**(3), 648–660 (2019)
18. Belotti, M., Božić, N., Pujolle, G., Secci, S.: A vademecum on blockchain technologies: when, which, and how. IEEE Commun. Surv. Tutor. **21**(4), 3796–3838 (2019)
19. Bergman, S., Asplund, M., Nadjm-Tehrani, S.: Permissioned blockchains and distributed databases: a performance study. Concurr. Comput.: Pract. Exp. **32**(12) (2020)
20. Bouachir, O., Aloqaily, M., Tseng, L., Boukerche, A.: Blockchain and fog computing for cyberphysical systems: the case of smart industry. Computer **53**(9), 36–45 (2020)
21. Brincat, A.A., Lombardo, A., Morabito, G., Quattropani, S.: On the use of Blockchain technologies in WiFi networks. Comput. Netw. **162**, 106,855 (2019)
22. Bui, T., Cooper, D., Collomosse, J., Bell, M., Green, A., Sheridan, J., Higgins, J., Das, A., Keller, J.R., Thereaux, O.: Tamper-proofing video with hierarchical attention autoencoder hashing on blockchain. IEEE Trans. Multimed. **22**(11), 2858–2872 (2020)
23. Bukhari, S.H.R., Rehmani, M.H., Siraj, S.: A survey of channel bonding for wireless networks and guidelines of channel bonding for futuristic cognitive radio sensor networks. IEEE Commun. Surv. Tutor. **18**(2), 924–948 (2016)
24. Caro, M.P., Ali, M.S., Vecchio, M., Giaffreda, R.: Blockchain-based traceability in agri-food supply chain management: a practical implementation. In: 2018 IoT Vertical and Topical Summit on Agriculture - Tuscany (IOT Tuscany), pp. 1–4 (2018)
25. Chandra, G.R., Liaqat, I.A., Sharma, B.: Blockchain redefining: the halal food sector. In: 2019 Amity International Conference on Artificial Intelligence (AICAI), pp. 349–354 (2019)
26. Dinh, T., Liu, R., Zhang, M., Chen, G., Ooi, B., Wang, J.: Untangling blockchain: a data processing view of blockchain systems. IEEE Trans. Knowl. Data Eng. **30**(07), 1366–1385 (2018)
27. Eisele, S., Barreto, C., Dubey, A., Koutsoukos, X., Eghtesad, T., Laszka, A., Mavridou, A.: Blockchains for transactive energy systems: opportunities, challenges, and approaches. Computer **53**(9), 66–76 (2020)
28. Elavarasi, G., Murugaboopathi, G., Kathirvel, S.: Fresh fruit supply chain sensing and transaction using IoT. In: 2019 IEEE International Conference on Intelligent Techniques in Control, Optimization and Signal Processing (INCOS), pp. 1–4 (2019)
29. Fan, C., Ghaemi, S., Khazaei, H., Musilek, P.: Performance evaluation of blockchain systems: a systematic survey. IEEE Access **8**, 126,927–126,950 (2020)
30. Ferdousi, T., Gruenbacher, D., Scoglio, C.M.: A permissioned distributed ledger for the us beef cattle supply chain. IEEE Access **8**, 154,833–154,847 (2020)
31. Fioretto, F., Mak, T.W.K., Van Hentenryck, P.: Differential privacy for power grid obfuscation. IEEE Trans. Smart Grid **11**(2), 1356–1366 (2020)
32. Gai, K., Wu, Y., Zhu, L., Qiu, M., Shen, M.: Privacy-preserving energy trading using consortium blockchain in smart grid. IEEE Trans. Ind. Inform. **15**(6), 3548–3558 (2019)
33. Ghimire, S., Choi, J.Y., Lee, B.: Using blockchain for improved video integrity verification. IEEE Trans. Multimed. **22**(1), 108–121 (2020)
34. Ghorbanian, M., Dolatabadi, S.H., Siano, P., Kouveliotis-Lysikatos, I., Hatziargyriou, N.D.: Methods for flexible management of blockchain-based cryptocurrencies in electricity markets and smart grids. IEEE Trans. Smart Grid **11**(5), 4227–4235 (2020)
35. Haroon, A., Basharat, M., Khattak, A.M., Ejaz, W.: Internet of things platform for transparency and traceability of food supply chain. In: 2019 IEEE 10th Annual Information Technology, Electronics and Mobile Communication Conference (IEMCON), pp. 0013–0019 (2019)
36. Hassan, M.U., Rehmani, M.H., Chen, J.: Privacy preservation in blockchain based IoT systems: integration issues, prospects, challenges, and future research directions. Futur. Gener. Comput. Syst. **97**, 512–529 (2019)

37. Hassan, M.U., Rehmani, M.H., Chen, J.: DEAL: differentially private auction for blockchain-based microgrids energy trading. IEEE Trans. Serv. Comput. **13**(2), 263–275 (2020)
38. Hassan, M.U., Rehmani, M.H., Chen, J.: Differential privacy in blockchain technology: a futuristic approach. J. Parallel Distrib. Comput. **145**, 50–74 (2020)
39. Hassan, M.U., Rehmani, M.H., Chen, J.: Differential privacy techniques for cyber physical systems: a survey. IEEE Commun. Surv. Tutor. **22**(1), 746–789 (2020)
40. Hassan, M.U., Rehmani, M.H., Chen, J.: Differentially private dynamic pricing for efficient demand response in smart grid. In: ICC 2020 - 2020 IEEE International Conference on Communications (ICC), pp. 1–6 (2020)
41. Hassan, M.U., Rehmani, M.H., Kotagiri, R., Zhang, J., Chen, J.: Differential privacy for renewable energy resources based smart metering. J. Parallel Distrib. Comput. **131**, 69–80 (2019)
42. Hebert, C., Di Cerbo, F.: Secure blockchain in the enterprise: a methodology. Pervasive Mob. Comput. **59**, 101,038 (2019)
43. Hiertz, G.R., Denteneer, D., Stibor, L., Zang, Y., Costa, X.P., Walke, B.: The IEEE 802.11 universe. IEEE Commun. Mag. **48**(1), 62–70 (2010)
44. Jiang, L., Xie, S., Maharjan, S., Zhang, Y.: Blockchain empowered wireless power transfer for green and secure internet of things. IEEE Netw. **33**(6), 164–171 (2019)
45. Juma, H., Shaalan, K., Kamel, I.: A survey on using blockchain in trade supply chain solutions. IEEE Access **7**, 184,115–184,132 (2019)
46. Kaleem, Z., Rehmani, M.H.: Amateur drone monitoring: state-of-the-art architectures, key enabling technologies, and future research directions. IEEE Wireless Commun. **25**(2), 150–159 (2018)
47. Kant, K., Pal, A.: Internet of perishable logistics. IEEE Internet Comput. **21**(1), 22–31 (2017)
48. Khan, A.A., Rehmani, M.H., Rachedi, A.: Cognitive-radio-based internet of things: applications, architectures, spectrum related functionalities, and future research directions. IEEE Wirel. Commun. **24**(3), 17–25 (2017)
49. Khan, A.A., Rehmani, M.H., Reisslein, M.: Cognitive radio for smart grids: survey of architectures, spectrum sensing mechanisms, and networking protocols. IEEE Commun. Surv. Tutor. **18**(1), 860–898 (2016)
50. Khan, A.A., Rehmani, M.H., Reisslein, M.: Requirements, design challenges, and review of routing and MAC protocols for CR-based smart grid systems. IEEE Commun. Mag. **55**(5), 206–215 (2017)
51. Koens, T., Poll, E.: Assessing interoperability solutions for distributed ledgers. Pervasive Mob. Comput. **59**, 101,079 (2019)
52. Kolokotronis, N., Limniotis, K., Shiaeles, S., Griffiths, R.: Secured by blockchain: safeguarding internet of things devices. IEEE Consum. Electron. Mag. **8**(3), 28–34 (2019)
53. Kotobi, K., Bilen, S.G.: Secure blockchains for dynamic spectrum access: a decentralized database in moving cognitive radio networks enhances security and user access. IEEE Veh. Technol. Mag. **13**(1), 32–39 (2018)
54. Kshetri, N.: Blockchain and the economics of food safety. IT Prof. **21**(3), 63–66 (2019)
55. Kshetri, N., DeFranco, J.: The economics behind food supply blockchains. Computer **53**(12), 106–110 (2020)
56. Kshetri, N., Loukoianova, E.: Blockchain adoption in supply chain networks in Asia. IT Prof. **21**(1), 11–15 (2019)
57. Kuo, P., Mourad, A., Ahn, J.: Potential applicability of distributed ledger to wireless networking technologies. IEEE Wirel. Commun. **25**(4), 4–6 (2018)
58. Kuzlu, M., Pipattanasomporn, M., Rahman, S.: Communication network requirements for major smart grid applications in HAN, NAN and WAN. Comput. Netw. **67**, 74–88 (2014)
59. Li, Z., Wu, H., King, B., Ben Miled, Z., Wassick, J., Tazelaar, J.: A hybrid blockchain ledger for supply chain visibility. In: 2018 17th International Symposium on Parallel and Distributed Computing (ISPDC), pp. 118–125 (2018)
60. Liu, Y., Yu, F.R., Li, X., Ji, H., Leung, V.C.M.: Blockchain and machine learning for communications and networking systems. IEEE Commun. Surv. Tutor. **22**(2), 1392–1431 (2020). DOI 1

61. Malik, S., Kanhere, S.S., Jurdak, R.: Productchain: Scalable blockchain framework to support provenance in supply chains. In: 2018 IEEE 17th International Symposium on Network Computing and Applications (NCA), pp. 1–10 (2018)
62. Mondal, S., Wijewardena, K.P., Karuppuswami, S., Kriti, N., Kumar, D., Chahal, P.: Blockchain inspired RFID-based information architecture for food supply chain. IEEE Internet Things J. **6**(3), 5803–5813 (2019)
63. Mouftah, H.T., Erol-Kantarci, M., Rehmani, M.H. (eds.): Transportation and Power Grid in Smart Cities: Communication Networks and Services. Wiley, Hoboken (2019)
64. Naeem, A., Rehmani, M.H., Saleem, Y., Rashid, I., Crespi, N.: Network coding in cognitive radio networks: a comprehensive survey. IEEE Commun. Surv. Tutor. **19**(3), 1945–1973 (2017)
65. Narayanaswami, C., Nooyi, R., Govindaswamy, S.R., Viswanathan, R.: Blockchain anchored supply chain automation. IBM J. Res. Dev. **63**(2/3), 7:1–7:11 (2019)
66. Nguyen, C.T., Hoang, D.T., Nguyen, D.N., Niyato, D., Nguyen, H.T., Dutkiewicz, E.: Proof-of-stake consensus mechanisms for future blockchain networks: fundamentals, applications and opportunities. IEEE Access **7**, 85727–85745 (2019)
67. Nguyen, D.C., Pathirana, P.N., Ding, M., Seneviratne, A.: Integration of blockchain and cloud of things: architecture, applications and challenges. IEEE Commun. Surv. Tutor. **22**(4), 2521–2549 (2020)
68. Ometov, A., Bardinova, Y., Afanasyeva, A., Masek, P., Zhidanov, K., Vanurin, S., Sayfullin, M., Shubina, V., Komarov, M., Bezzateev, S.: An overview on blockchain for smartphones: state-of-the-art, consensus, implementation, challenges and future trends. IEEE Access **8**, 103,994–104,015 (2020)
69. Pal, A., Kant, K.: Iot-based sensing and communications infrastructure for the fresh food supply chain. Computer **51**(2), 76–80 (2018)
70. Pal, A., Kant, K.: Using blockchain for provenance and traceability in internet of things-integrated food logistics. Computer **52**(12), 94–98 (2019)
71. Panda, S.K., Blome, A., Wisniewski, L., Meyer, A.: Iot retrofitting approach for the food industry. In: 2019 24th IEEE International Conference on Emerging Technologies and Factory Automation (ETFA), pp. 1639–1642 (2019)
72. Refaey, A., Hammad, K., Magierowski, S., Hossain, E.: A blockchain policy and charging control framework for roaming in cellular networks. IEEE Netw. **34**(3), 170–177 (2020)
73. Rehan, M.M., Rehmani, M.H. (eds.): Blockchain-Enabled Fog and Edge Computing: Concepts, Architectures and Applications. CRC Press, Boca Raton (2020)
74. Rehmani, M.H., Akhtar, F., Davy, A., Jennings, B.: Achieving resilience in SDN-based smart grid: a multi-armed bandit approach. In: 2018 4th IEEE Conference on Network Softwarization and Workshops (NetSoft), pp. 366–371 (2018)
75. Rehmani, M.H., Davy, A., Jennings, B., Assi, C.: Software defined networks-based smart grid communication: a comprehensive survey. IEEE Commun. Surv. Tutor. **21**(3), 2637–2670 (2019)
76. Rehmani, M.H., Dhaou, R. (eds.): Cognitive Radio, Mobile Communications and Wireless Networks. EAI/Springer Innovations in Communications and Computing book series (2019)
77. Rehmani, M.H., Pathan, A.S.K. (eds.): Emerging Communication Technologies Based on Wireless Sensor Networks: Current Research and Future Applications. CRC Press, Taylor and Francis Group, New York (2016)
78. Rehmani, M.H., Reisslein, M., Rachedi, A., Erol-Kantarci, M., Radenkovic, M.: Integrating renewable energy resources into the smart grid: recent developments in information and communication technologies. IEEE Trans. Ind. Inform. **14**(7), 2814–2825 (2018)
79. Saleem, Y., Crespi, N., Rehmani, M.H., Copeland, R.: Internet of things-aided smart grid: technologies, architectures, applications, prototypes, and future research directions. IEEE Access **7**, 62962–63003 (2019)
80. Saleem, Y., Yau, K.A., Mohamad, H., Ramli, N., Rehmani, M.H., Ni, Q.: Clustering and reinforcement-learning-based routing for cognitive radio networks. IEEE Wirel. Commun. **24**(4), 146–151 (2017)

81. Sedjelmaci, S., Brahmi, I.H., Ansari, N., Rehmani, M.H.: Cyber security framework for vehicular network based on a hierarchical game. IEEE Trans. Emerg. Top. Comput. 1–1 (2018)
82. Sengul, C.: Distributed ledgers for spectrum authorization. IEEE Internet Comput. **24**(3), 7–18 (2020)
83. Siano, P., De Marco, G., Rolán, A., Loia, V.: A survey and evaluation of the potentials of distributed ledger technology for peer-to-peer transactive energy exchanges in local energy markets. IEEE Syst. J. **13**(3), 3454–3466 (2019)
84. Tsang, Y.P., Choy, K.L., Wu, C.H., Ho, G.T.S., Lam, H.Y.: Blockchain-driven IoT for food traceability with an integrated consensus mechanism. IEEE Access **7**, 129,000–129,017 (2019)
85. Tushar, W., Saha, T.K., Yuen, C., Smith, D., Poor, H.V.: Peer-to-peer trading in electricity networks: an overview. IEEE Trans. Smart Grid **11**(4), 3185–3200 (2020)
86. Ul Hassan, N., Yuen, C., Niyato, D.: Blockchain technologies for smart energy systems: fundamentals, challenges, and solutions. IEEE Ind. Electron. Mag. **13**(4), 106–118 (2019)
87. Uriarte, R.B., DeNicola, R.: Blockchain-based decentralized cloud/fog solutions: challenges, opportunities, and standards. IEEE Commun. Stand. Mag. **2**(3), 22–28 (2018)
88. Usmani, M.M.T.: An Introduction to Islamic Finance. Maktaba Ma'ariful Quran - Quranic Studies Publishers (2002)
89. Wang, F., Chen, Y., Wang, R., Francis, A.O., Emmanuel, B., Zheng, W., Chen, J.: An experimental investigation into the hash functions used in blockchains. IEEE Trans. Eng. Manag. **67**(4), 1404–1424 (2020)
90. Wang, S., Li, D., Zhang, Y., Chen, J.: Smart contract-based product traceability system in the supply chain scenario. IEEE Access **7**, 115,122–115,133 (2019)
91. Wang, T., Zheng, Z., Rehmani, M.H., Yao, S., Huo, Z.: Privacy preservation in big data from the communication perspective-a survey. IEEE Commun. Surv. Tutor. **21**(1), 753–778 (2019)
92. Wang, W., Hoang, D.T., Hu, P., Xiong, Z., Niyato, D., Wang, P., Wen, Y., Kim, D.I.: A survey on consensus mechanisms and mining strategy management in blockchain networks. IEEE Access **7**, 22328–22370 (2019)
93. Weiss, M.B.H., Werbach, K., Sicker, D.C., Bastidas, C.E.C.: On the application of blockchains to spectrum management. IEEE Trans. Cogn. Commun. Netw. **5**(2), 193–205 (2019)
94. Wright, D.: The history of the IEEE 802 standard. IEEE Commun. Stand. Mag. **2**(2), 4–4 (2018)
95. Wüst, K., Gervais, A.: Do you need a blockchain? In: Crypto Valley Conference on Blockchain Technology (CVCBT), pp. 45–54 (2018)
96. Wu, C.K., Tsang, K.F., Liu, Y., Zhu, H., Wei, Y., Wang, H., Yu, T.T.: Supply chain of things: a connected solution to enhance supply chain productivity. IEEE Commun. Mag. **57**(8), 78–83 (2019)
97. Xiao, Y., Zhang, N., Lou, W., Hou, Y.T.: A survey of distributed consensus protocols for blockchain networks. IEEE Commun. Surv. Tutor. **22**(2), 1432–1465 (2020)
98. Xie, J., Yu, F.R., Huang, T., Xie, R., Liu, J., Liu, Y.: A survey on the scalability of blockchain systems. IEEE Netw. **33**(5), 166–173 (2019)
99. Xiong, Z., Zhang, Y., Niyato, D., Wang, P., Han, Z.: When mobile blockchain meets edge computing. IEEE Commun. Mag. **56**(8), 33–39 (2018)
100. Yang, X., Wang, G., He, H., Lu, J., Zhang, Y.: Automated demand response framework in ELNs: decentralized scheduling and smart contract. IEEE Trans. Syst. Man Cybern.: Syst. **50**(1), 58–72 (2020)
101. Yang, X., Wang, T., Ren, X., Yu, W.: Survey on improving data utility in differentially private sequential data publishing. IEEE Trans. Big Data 1–1 (2017)
102. Yu, F.R., Liu, J., He, Y., Si, P., Zhang, Y.: Virtualization for distributed ledger technology (vDLT). IEEE Access **6**, 25019–25028 (2018)
103. Zhang, X., Sun, P., Xu, J., Wang, X., Yu, J., Zhao, Z., Dong, Y.: Blockchain-based safety management system for the grain supply chain. IEEE Access **8**, 36398–36410 (2020)

104. Zhou, Z., Wang, B., Dong, M., Ota, K.: Secure and efficient vehicle-to-grid energy trading in cyber physical systems: integration of blockchain and edge computing. IEEE Trans. Syst. Man Cybern.: Syst. **50**(1), 43–57 (2020)
105. Zhu, T., Li, G., Zhou, W., Yu, P.S.: Differentially private data publishing and analysis: a survey. IEEE Trans. Knowl. Data Eng. **29**(8), 1619–1638 (2017)

Index

Symbols
2G, 114
3G, 114
4G, 114
51% attack, 61
5G, 112
6G, 112

A
Access control, 124
Access method, 120
Acronyms, list of, xv
Adler32, 86
Aldi, 11
Algorand, 76, 85
Anonymity, 26
Application layer, 30
Arbitrage, 124
Asset, 10, 55
Auction mechanism, 116
Auditability, 10
Auto-mobile industry, 81
Availability, 9

B
Base Transceiver Station (BTS), 114
Bitcoin, 26, 38, 39, 41, 56, 142
Block, 6
BlockBench, 46
Block body, 41
Blockchain, 5, 10
Blockchain as a Service (BaaS), 38
Block confirmation, 34
Block header, 40

Block size, 26
Block time, 70
Broadcasting, 27
Business automation, 10
Byteball, 7
Byzantine Fault Tolerant (BFT), 62, 77

C
Cardano, 75
Carpooling, 124
Casper, 76, 85
Cellular network, 114, 124, 134
Central server, 4
Chain Core, 38, 44
Chain of activity, 85
Channel hopping, 124
Citizen Band Radio Service (CBRS), 117
Claude Shannon, 107
Client-server model, 5, 23
Cloud computing, 23
Cloud of Things (CoT), 132
Code-bases, 37
Cognitive Radio (CR), 109
Cognitive radio networks, 116
Collision free communication, 124
Committee-based consensus protocol, 74
Communication, 107
Communication networks, 107, 123
Computational power, 63
Computing, 23, 132
Confidentiality, 36
Consensus, 19, 39, 61, 62
Consensus layer, 31
Consortium blockchain, 18, 25

© Springer Nature Switzerland AG 2021
M. H. Rehmani, *Blockchain Systems and Communication Networks: From Concepts to Implementation*, Textbooks in Telecommunication Engineering,
https://doi.org/10.1007/978-3-030-71788-9

Consumer Electronics (CE) devices, 128
Contract account, 42
Corda, 7, 42
COVID-19, 107
CPU, 32
CR nodes, 116
CRC32, 86
Credit card, 142
Cryptocurrency, 6, 41, 53, 68, 142
Cryptograhic puzzle, 64
Cryptographic, 6
Cryptographic puzzle, 77
Cyber attacks, 130

D
D2D networks, 117
DAGbench, 47
Data model, 45
Data organization and topology layer, 32
Data roaming, 134
Data sharing, 124, 131
Data storage, 21
Data tampering, 10
Database, 4, 15
Database Management System (DBMS), 15
Database systems, 8
DDoS attack, 142
Decentralization, 8
Decentralized, 23
Decision tree, 145
Deployment, 36
Differential privacy, 139
Difficulty, 66
Digital currency, 6
Digital ledgers, 3
Digital signature, 45
Directed Acyclic Graph (DAG), 5, 7
Dissemination, 27
Distributed, 23
Distributed database, 61
Distributed digital ledgers, 4
Distributed Hash Table (DHT), 27, 42
Distributed Ledger (DL), 4, 123
Distributed Ledger Technology (DLT), 4
DNS server, 27
DoS attack, 39
Double spending problem, 65
Dynamic Spectrum Access (DSA), 109

E
Electricity, 142

Electric vehicles, 141
Energy consumption, 25
Energy trading, 81, 124, 138
Ethash, 48
Ethereum, 31, 38, 41, 56, 142
Ethereum owned account, 42

F
Federal   Communication   Commission
       (FCC), 109
Federated learning, 146
FIAT currency, 55
Fiat currency, 112
Filecoin, 74
Fog-RAN, 126
Fog/edge computing, 23
Food supply chain, 11
Frequency band, 110
FTS algorithm, 76
Full blockchain nodes, 28

G
Genesis block, 40
GHOST, 49
Giga hertz (GHz), 108
Github, 37
Global System for Mobile Communication
       (GSM), 107
GPU, 32
Grant free access, 124
Grinding attack, 75

H
Hardware, 132
Hardware layer, 32
Hash, 41
Hash pointer, 35
Hashing, 19, 47, 85
Hashing function, 47
Hashlib, 85
Heath and Safety Authority, 11
Home Network Mobile Operator (HNMO),
       134
Hyperledger, 42
Hyperledger Caliper, 46
Hyperledger Fabric, 31, 38

I
Identity of nodes, 4
Immutability, 9, 20, 47

Industrial, Scientific, and Medical (ISM)
    band, 109
Information, 27
Infrastructure, 38
Internet, 23, 107
Interoperability, 52
IoT network, 125
IOTA, 7

J
Java Virtual Machine (JVM), 42

K
Kademlia, 27
Kopercoin, 74

L
Latency, 65
Leader node, 64
Ledger, 3, 15, 120
Ledger notification, 36
Ledger storage, 128
Licensed band, 113
Lidl, 11
Lightweight blockchain nodes, 28
Limited block size, 25
Long-range attack, 75

M
Machine learning, 144
Magic number, 40
Malicious users, 138
Markov decision process, 145
Medical image analysis, 144
Merkle Tree, 35
Merkle Tree root, 41
Mined, 34
Miners, 17
Mini project, 81
Mining pool, 68
Mining process, 25
Mining puzzle, 66
Mining reward, 65
Mobile devices, 120, 133
Mobile network operator, 118
Multi Armed Bandit (MAB), 145
Multimedia, 134
MySQL, 16

N
Nano, 7
Native tokens, 56
Negative energy pricing, 143
Netfliz, 135
Network layer, 31
Networking, 4, 132
Neural networks, 145
Non-native tokens, 56
Non-repudiation, 9
Nothing-at-stake attack, 75

O
OmmersHash, 48
On-chain data management, 48, 49
Open-source, 37
Oracle, 16
Ordered transactions, 6
Ouroboros, 85
Ouroborous, 75
Ownership, 6

P
P2P energy trading system, 147
P2P networking, 5
Pakistan, 3
Parent block, 48
ParentHash, 48
Parties, 10
Patwari, 3
Payroll, 3
PCs, 4
Peer discovery, 27
Peersensus, 74
Performance evaluation, 45
Permicoin, 75
Permissioned, 4
Permissioned blockchain, 25
Permissionless, 4
Permissionless blockchain, 25
Pharmaceutical industry, 81
PR activity, 116
PR nodes, 116
Previous block, 41
Primary network, 118
Privacy preservation, 139
Private, 4
Private blockchain, 18, 24, 42
Private key, 29
Proof of authority, 42
Proof-of-Exercise, 75

Proof-of-Human-Work, 74
Proof-of-Stake (PoS), 67
Proof-of-Useful-Work, 75
Proof-of-Work (PoW), 39, 63, 65
Pseudonymity, 9, 25, 77
Public, 4
Public blockchain, 18, 24, 26
Public key, 19
Public verifiability, 19
Python, 85

Q
Q-learning, 145

R
R3, 42
Radio Access Network (RAN), 126
Radix, 7
Real estate, 81
Receive node, 29
Reinforcement learning, 144
Riding service, 131
Roaming, 124, 134

S
Satellite communication, 117
Scalability, 25, 26, 50
Secondary network, 118
Security, 9
Sender node, 29
Service providers, 38
Sharding, 52
Shared licensed band, 113
Smart contract, 31, 118
Smart factory, 81
Smart grid, 107, 138
Smart meter, 140
Software as a Service (SaaS), 38
Solidity, 31
Sp8de, 75
SPECTRE, 49
Spectrum, 109
Spectrum auction, 124
Spectrum mobility, 123
Spectrum patrolling, 124
Spectrum regulation and auditing, 124
Spectrum regulator, 117
Spectrum scarcity, 110
Spectrum selection, 123
Spectrum sensing, 123
Spectrum sensing as a service, 124

Spectrum sharing, 110, 123
Spectrum utilization, 116
Speech recognition, 144
Stake, 65
Storage, 23, 132
SuperValu, 11
Supply chain, 81
Surplus energy, 130
Sybil attacks, 26
Symbols, list of, xv

T
Tampered, 6
Tendermint, 43, 77, 85
Tera hertz (THz), 108
Tesco, 11
Third-party, 5, 53
Throughput, 85
Tokens, 7
Topology maintenance, 27
Traditional networks, 133
Transaction broadcast, 34
Transaction handling capacity, 51
Transaction per second, 70
Transaction per second (tps), 42, 55
Transaction processing time, 55
Transaction signing, 34
Transaction speed, 25
Transaction validation, 34
Transaction verification, 34
Transactions, 6, 33
Transactive energy system, 138
Transparency, 9
Trust, 21
Trusted third-party, 131
TV White Space, 117

U
Uncle block, 48
Under-utilized spectrum, 116
Unlicensed spectrum, 114
Users, 27
UTXO, 39

V
Validation, 41
Validator nodes, 29
VANETs, 131
Version, 41
Video integrity, 135
Video streaming service, 135

Virtual machine, 31
Virtual power plants (VPPs), 138
Virtualization layer, 31
Vp2p, 42
Vulnerability, 26

W
Wi-Fi, 117

Wire-based communication system, 108
Wireless based communication system, 108
Wireless network, 125
Wireless power transfer, 124
Wireless radio spectrum, 109

X
X.509, 42

Printed in the United States
by Baker & Taylor Publisher Services